PROCESS MODELING

Morton M. Denn

Department of Chemical Engineering,
University of California, Berkeley

Longman
New York & London

Pitman Publishing Inc.
1020 Plain Street
Marshfield, Massachusetts 02050

PROCESS MODELING

Longman Inc., 95 Church Street, White Plains, N.Y. 10601
Associated companies, branches and representatives throughout the world.

Copyright © 1986 by Morton M. Denn

All rights reserved. No part of this publication may be reproduced, stored in a retrieval system, or transmitted in any form or by any means, electronic, mechanical, photocopying, recording, or otherwise, without the prior permission of the publisher.

Library of Congress Cataloging in Publication Data

Denn, Morton M., 1939–
 Process modeling.

 Includes bibliographies and index.
 1. Chemical processes—Mathematical models.
I. Title.
TP155.7.D46   1985   660.2′8′00724   85-9393
ISBN 0-273-08704-5

Manufactured in the United States of America
Printing: 9 8 7 6 5 4 3 2 1   Year: 94 93 92 91 90 89 88 87 86

CONTENTS

Preface   x

1
WHAT IS A MODEL?   *1*

1.1   INTRODUCTION   1
1.2   EXAMPLE: A MECHANICAL SYSTEM   2
    1.2.1   Fundamental   3
    1.2.2   Empirical   4
    1.2.3   Analogy   5
1.3   RESERVATIONS AND LIMITS   7
1.4   A LOGICAL HIERARCHY   9
BIBLIOGRAPHICAL NOTES   10

2
WHY MODEL?   *12*

2.1   INTRODUCTION   12
2.2   SETTING THE STATE   15
2.3   FLUID CATALYTIC CRACKER (FCC)   15
2.4   MOVING BED COAL GASIFIER   17
2.5   FIBER SPINLINE   24
2.6   WASTEWATER TREATMENT   28
BIBLIOGRAPHICAL NOTES   30

3
CONSTRUCTING A MODEL — 31

3.1	INTRODUCTION	31
3.2	VARIABLES	31
3.3	A WELL-STIRRED TANK	32
3.4	MODELING LOGIC	37
3.5	CONSTITUTIVE EQUATIONS	37
3.6	CONTROL VOLUME	40
3.7	REPRISE: DENSITY OF LIQUID SYSTEMS	41
	BIBLIOGRAPHICAL NOTES	43

4
MODELING REACTORS: CONSERVATION OF MASS — 45

4.1	INTRODUCTION	45
4.2	CONSERVATION OF MASS	46
4.3	A DESIGN EXAMPLE	49
4.4	TRANSIENTS AND TIME CONSTANTS	52
4.5	CATALYTIC OXIDATION: MULTIPLICITY AND INSTABILITY	53
4.6	BIOLOGICAL AND OTHER AUTOCATALYTIC SYSTEMS	57
4.7	BATCH AND TUBULAR REACTORS	60

5
MODELING REACTORS: CONSERVATION OF ENERGY — 62

5.1	INTRODUCTION	62
5.2	THERMODYNAMIC VARIABLES	62
5.3	CONSERVATION OF ENERGY	65
5.4	BATCH AND TUBULAR REACTORS	71
5.5	A CLASSIC (ALL TOO COMMON) BLUNDER	72
5.6	ORDER REDUCTION	73
5.7	MULTIPLICITY	74
5.8	TRANSIENT BEHAVIOR	81

5.9	FLUID CATALYTIC CRACKER (FCC) MODEL	84
	5.9.1 The Problem	85
	5.9.2 Model Equations	86
	5.9.3 Validation	90
	5.9.4 Multiplicity	90
	5.9.5 Order Reduction	93
	5.9.6 Application	94
	5.9.7 Final Remarks	96
BIBLIOGRAPHICAL NOTES		98

6

SCALING 103

6.1	INTRODUCTION	103
6.2	THE ENERGY BALANCE	103
6.3	THE HEAT EQUATION	107
6.4	SCALING THE HEAT EQUATION	109
6.5	APPLYING THE EINSTEIN TIME/LENGTH CONCEPT	111
6.6	TRANSPORT OF MASS, MOMENTUM, AND ENERGY	112
	6.6.1 Problem Statement	112
	6.6.2 Fluid Flow	113
	6.6.3 Heat Transfer	118
	6.6.4 Some Conclusions	120
6.7	A LOGICAL PROBLEM	120
6.8	SENSITIVITY	122
6.9	CONCLUDING REMARKS	123
BIBLIOGRAPHICAL NOTES		124

7

CASE STUDY: COAL GASIFICATION 125

7.1	INTRODUCTION	125
7.2	OVERALL CONSTRAINTS	126
7.3	KINETICS-FREE MODELING	131
7.4	DETAILED MODELING	136
	7.4.1 Broad Principles	136
	7.4.2 One-Dimensional Homogeneous Model	138
	7.4.3 Model Results	143
	7.4.4 Engineering Application	143

	7.4.5 Two-Dimensional Homogeneous Model	148
	7.4.6 Transient Homogeneous Model	156
7.5	SENSITIVITY	158
7.6	CONCLUDING REMARKS	160
	BIBLIOGRAPHICAL NOTES	160

8
STABILITY 162

8.1	INTRODUCTION	162
8.2	LIAPUNOV FUNCTION	163
8.3	STABILITY TO INFINITESIMAL PERTURBATIONS	166
8.4	CONTINUOUS-FLOW STIRRED REACTOR: I	167
8.5	CONTINUOUS-FLOW STIRRED REACTOR: II	169
8.6	FINITE STABILITY REGIONS FROM LINEAR THEORY	172
8.7	CONCLUDING REMARKS	176
	BIBLIOGRAPHICAL NOTES	177

9
MODAL ANALYSIS 178

9.1	INTRODUCTION	178
9.2	LINEARIZATION	178
9.3	EIGENVALUES AND EIGENVECTORS	180
9.4	NON-HOMOGENEOUS SYSTEMS	183
9.5	ORDER REDUCTION	183
9.6	MODAL CONTROL	184
9.7	FLUID CATALYTIC CRACKER	185
	9.7.1 Linear Model	185
	9.7.2 Control	187
	9.7.3 Measurement and Response	192
9.8	SLIGHTLY NONLINEAR PROCESSES	194
	9.8.1 Quadratic Expansion	194
	9.8.2 Linearization and Stability Limits	196
	9.8.3 Batch Fluidized Bed	196
9.9	CONCLUDING REMARKS	203
	BIBLIOGRAPHICAL NOTES	204

10

DISCRETIZATION — 205

10.1	INTRODUCTION	205
10.2	METHODS OF WEIGHTED RESIDUALS	205
10.3	THE ADJOINT OPERATOR	209
10.4	CALCULUS OF VARIATIONS	211
10.5	RITZ-GALERKIN METHOD	214
10.6	FINITE-ELEMENT METHODS	215
10.7	CONCLUDING REMARKS	216
BIBLIOGRAPHICAL NOTES		217

11

FROM DISCRETE TO CONTINUOUS — 219

11.1	INTRODUCTION	219
11.2	BINARY DISTILLATION	220
11.3	CONTINUOUS APPROXIMATION	223
11.4	BATCH POLYMERIZATION	225
11.5	AN ALTERNATIVE FORMALISM	227
BIBLIOGRAPHICAL NOTES		228

12

CASE STUDY: FIBER SPINLINE — 230

12.1	INTRODUCTION		230
12.2	THIN FILAMENT EQUATIONS		231
	12.2.1	Problem Definition	231
	12.2.2	Derivation	232
	12.2.3	Stress Constitutive Equation	236
	12.2.4	Phenomenological Coefficients	240
	12.2.5	Boundary Conditions	241
12.3	ASYMPTOTIC SOLUTIONS		243
	12.3.1	Introduction	243
	12.3.2	Isothermal, Low-Speed Steady Spinning	243
	12.3.3	Temperature Profile	246

12.4	FINITE-ELEMENT ANALYSIS	246
12.5	STEADY-STATE SPINLINE SIMULATION	248
	12.5.1 Pilot Plant Results	248
	12.5.2 Parametric Sensitivity	250
	12.5.3 Possible Plant Application	255
	12.5.4 Some Numerical Experiments	256
12.6	DRAW RESONANCE	260
12.7	DYNAMIC SENSITIVITY	265
12.8	CONCLUDING REMARKS	266
BIBLIOGRAPHICAL NOTES		267

13

MODEL VALIDATION 270

13.1	INTRODUCTION	270
13.2	STRUCTURAL SENSITIVITY AND TOVES	270
13.3	COAL GASIFIER: SIMPLICITY	271
13.4	FIBER SPINLINE: SIMPLICITY PARADOX	272
13.5	DEGREE OF FALSIFIABILITY	274
13.6	CONCLUDING REMARKS	275
BIBLIOGRAPHICAL NOTES		275

14

CASE STUDY: ACTIVATED SLUDGE PROCESS 277

14.1	INTRODUCTION	277
14.2	PROCESS CONDITIONS	279
14.3	PROCESS ECONOMICS	280
14.4	BASE CASE	282
14.5	DESIGN SENSITIVITY	285
14.6	DYNAMICS AND CONTROL	290
	14.6.1 Recycle Systems	290
	14.6.2 Plant Dynamics	295
	14.6.3 Oxygen Control	297
14.7	CONCLUDING REMARKS	300
APPENDIX TO CHAPTER 14		
14.A1	BIOCHEMICAL REACTOR (AERATION BASIN)	302

14.A2	CONTINUOUS SEDIMENTATION	305
	14.A2.1 Introduction	305
	14.A2.2 Flux Theory	307
	14.A2.3 Settling Velocity	311
	14.A2.4 Clarification	313
	14.A2.5 Design	313
	14.A2.6 Dynamics	314
14.A3	PRIMARY CLARIFIER	316
14.A4	OTHER UNIT OPERATIONS	317
BIBLIOGRAPHICAL NOTES		318

A Final Comment 320

Index 321

PREFACE

Process Modeling is my attempt to present a systematic treatment of the methods that are useful in the development and application of mathematical models. It is not a text in applied mathematics, although mathematics must always be used in the interpretation and application of a model. Implementation of a model (in the form of a computer code for simulation, for example) is usually straightforward, though perhaps tedious; the truly challenging aspect of modeling is in the use of physical principles to arrive at a proper mathematical formulation. (Indeed, the engineering literature contains too many examples of models in which the basic equations are incorrect, some in commercially offered computer codes. This point is discussed in some detail in Chapter 5, where the Bibliographical Notes contain a sampling of important textbooks and research papers with fundamental errors in basic equations.)

The present book began to take form in the outline of a graduate course that I taught in 1979 at the Technion-Israel Institute of Technology, which I have since repeated at the University of California, Berkeley. The overall approach dates to a collaboration at the University of Delaware with T. W. F. Russell, which culminated in the publication of our text *Introduction to Chemical Engineering Analysis* (1972); the intent and level of that text are quite different from the present one, but the treatment in *Process Modeling* owes much to the conceptual framework of *Introduction*.

An overview of this type is of necessity a personal one, and while I am indebted to coworkers who have contributed to my understanding of various aspects of the subject, they must be absolved of any responsibility for the way in which I have used their knowledge and insight. Jude Wallace introduced me to the philosophy-of-science literature relevant to model validation (Chapter 13). The three major case studies, coal gasification (Chapter 7), fiber spinning (Chapter 12), and activated sludge wastewater treatment (Chapter 14), summarize significant collaborations with many colleagues and students; these collaborations will be evident through the Bibliographical Notes, but it is appropriate that I here acknowledge in particular Robert Paterson, Reuel Shinnar, and James Wei.

This book was written during a time of many changes for me, and I appreciate the understanding and support of William Roberts of Pitman, who encouraged the start of this project with reasonable but unfulfilled expectations of the duration of the task. Gina Whitney contributed far more to the completion than Fig. 7.5. Portions of the text were drafted during periods in residence at the University of Melbourne and the California Institute of Technology, and the largest part was written during my inaugural years at Berkeley; my understanding of process modeling developed during my seventeen years at the University of Delaware, however, and I am grateful for the opportunities afforded by that institution. The debt to Fraser Russell will be apparent to anyone familiar with the clarity of his approach to problems in engineering and science; I doubt that the opportunity to write this book would have developed had I not had the fortune to work with him early in my career, and I am pleased to dedicate *Process Modeling* to him in token repayment.

Morton M. Denn

Melbourne, November 1985.

1

WHAT IS A MODEL?

1.1 INTRODUCTION

Mathematical model is a phrase that permeates the literature of modern engineering and science. This widespread usage has resulted in a range of meanings, and there is considerable confusion as to what a model is in any given context.* The term "modeler" is an epithet to many engineering practitioners; to others, mathematical modeling is simply synonymous with good engineering practice.

As we shall use the term here, a model is a quantification of a physical process that enables prediction of process behavior. More precisely,

> *A mathematical model of a process is a system of equations whose solution, given specified input data, is representative of the response of the process to a corresponding set of inputs.*

The "process" might be physical, biological, social, etc. (Students in our modeling courses have written papers on topics ranging from fluid catalytic

---

***The American Heritage Dictionary of the English Language* lists seven meanings of the noun *model;* two are relevant to our discussion: "(1) A small object . . . that represents some existing object" and "(2) A tentative ideational [i.e., conceptual] structure used as a testing device . . ." Scientific usage embodies both of these meanings, but neither is adequate. See the books cited in the Bibliographical Notes for a detailed discussion of the range of meanings that may be encountered.

1

cracking of petroleum to human learning and the U.S. economy, for example.) Our emphasis here will be on physical systems of industrial type, because such systems best elucidate the fundamental principles that we believe to be important in modeling. Several such systems, which will serve as useful examples from time to time, are described in subsequent sections; these include a reactor for converting coal to clean gaseous products, a spinline for manufacturing synthetic fibers, and a system for biological treatment of municipal wastewater.

It is preferable at this point to remain somewhat vague as to the manner in which we use "representative," since this concept cannot be separated from the question "why model?" which is discussed subsequently. We presume that we can identify each variable appearing in the equations of the model with an entity associated with the process; each entity must be measurable, at least in principle. (Students often question the insistence on measurable quantities. Measurement in principle is equivalent to insistence that any definition be an *operational definition;* that is, the definition must contain within itself the description of an experiment for measuring the defined quantity. The qualification *in principle* resolves any objection regarding experimental cost, or even the availability of appropriate equipment. A quantity that can never be measured in principle has no physical meaning.)

Scientists and engineers use (and intermix) three identifiable methodologies to obtain the equations for a mathematical model. These can be categorized as follows:

1. *Fundamental:* Use the accepted theory of the underlying sciences to derive equations. In this case, *the theories that we accept are our basic axioms in the logical process of model construction.*
2. *Empirical:* Use direct observations to develop equations that describe the experiments.
3. *Analogy:* Use the equations describing a system believed to be analogous, with variables identified by analogy on a one-to-one basis.

Each of these approaches is sketched out in the following section for an elementary system that is well understood. We will then return to the detailed examination of the questions of "why?" and "how?"

1.2 EXAMPLE: A MECHANICAL SYSTEM

The mass-spring-dashpot system shown in Fig. 1.1 is a mechanical system that is generally familiar from introductory physics courses, and it therefore provides a useful example of the several approaches to modeling. An ex-

1.2 EXAMPLE: A MECHANICAL SYSTEM

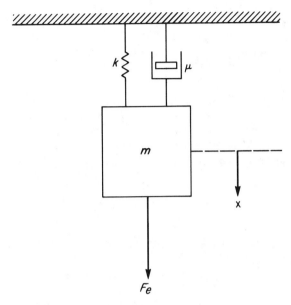

FIG. 1.1.
Mass-spring-dashpot system.

ternal force F_e is applied to the body of mass m, which is connected through parallel linkages to a rigid support; one linkage contains a spring, the other a dashpot (a shock absorber). The motion of the body induced by the external force is one-dimensional, and displacement from an arbitrary origin is denoted by x, with the positive x-direction as shown on the figure; we wish to be able to predict the position $x(t)$ of the body as a function of time for an arbitrary time-dependent force $F_e(t)$.

1.2.1 Fundamental

We take Newton's Second Principle as given: *The rate of change of momentum of the body equals the sum of the imposed forces.* Three colinear forces act on the body: the imposed force, F_e; the force of the linkage containing the spring, denoted F_s; and the force of the linkage containing the dashpot, denoted F_d. If gravity acts on the body, then the gravitational force will be included in F_e. Momentum is $m\dot{x}$. (The dot ˙ denotes the time derivative, d/dt; the notation is Newton's, and is thus particularly appropriate here.) For constant mass m, we can then write Newton's Second Principle as

$$m\ddot{x} = F_e + F_s + F_d \tag{1.1}$$

Hooke's law for a spring states that the force is proportional to the extension; if we take the coordinate origin as the position of the center of mass when the spring is unextended, then we may write

$$F_s = -kx \qquad (1.2)$$

The parameter k is the spring constant. The algebraic sign is negative because the force acts on the body in the negative direction.

If the dashpot is filled with a Newtonian oil, then the damping force is proportional to the velocity, and acts in the direction opposite to the motion:

$$F_d = -\mu \dot{x} \qquad (1.3)$$

μ is the coefficient of viscous damping. We thus obtain the model equation

$$m\ddot{x} + \mu \dot{x} + kx = F_e(t) \qquad (1.4)$$

The solution of this equation for arbitrary functions $F_e(t)$ is straightforward, given initial values of x and $\dot{x}$. The parameters m, μ, and k must be known or measured. k can be determined in a static experiment (i.e., $\dot{x} = \ddot{x} = 0$) under constant load: $k = F_e/x$; μ can be measured only dynamically, either on this system or in an independent experiment.

It is very important to note that Eqs. (1.2) and (1.3) are not consequences of Newton's second principle; they represent independent postulates. The momentum principle is always applicable, as long as we are dealing with nonrelativistic phenomena (which will always be the case here), but we know that not all springs satisfy Hooke's law, nor are all dashpots linear. There is clearly, then, a hierarchy, in which certain of the principles employed are more fundamental than others. This observation is central to the art of mathematical modeling, and we shall return to it later for more detailed consideration.

1.2.2 Empirical

An empirical model is simply an equation that records the relationship between system inputs and outputs (F_e and x, respectively, in this case). Many procedures exist for determining the appropriate experimental design, and the way in which data are converted to an equation. Statistical regression is a frequent source of such models. It is usually necessary to make some *a priori* assumptions about the mathematical structure of the equations. We

1.2 EXAMPLE: A MECHANICAL SYSTEM

shall give one example, recognizing that this is by no means a complete or even a representative treatment.

Most procedures for constructing empirical models of dynamical systems require the assumption that the model is linear:

$$a_n x^{(n)} + a_{n-1} x^{(n-1)} + \cdots + a_1 \dot{x} + a_0 x$$
$$= b_m f^{(m)} + b_{m-1} f^{(m-1)} + \cdots + b_1 \dot{f} + b_0 f \quad (1.5)$$

The *frequency response* method estimates the order of the dynamics (n and m) and the coefficients $\{a_i\}$, $\{b_i\}$ by making use of the following property of Eq. (1.5) with constant coefficients: if $f(t) = A \cos \omega t$, then $x(t) = B \cos(\omega t + \phi)$. The amplitude ratio, B/A, and the phase angle, ϕ, are functions of frequency, ω, and the shapes of these functions depend on order and coefficient values. Figure 1.2 is a plot (known as a *Bode diagram*) for a second-order system of the form

$$\tau^2 \ddot{x} + 2\zeta\tau \dot{x} + x = f(t) \quad (1.6)$$

Log (amplitude ratio) drops off with the -2 power of log ω at high frequency, and ϕ goes asymptotically to $-180°$; the more general relations are $-(n - m)$ power and $-(n - m) \times 90°$ respectively. The magnitude of the resonant peak near $\omega\tau = 1$ depends on the damping coefficient, ζ, and a peak occurs only for $\zeta^2 > 0.5$. The Bode diagram for the mass-spring-dashpot system would presumably have the form of Fig. 1.2, with the asymptotic value of the amplitude ratio for $\omega \to 0$ defining the scale factor between $f(t)$ and $F_e(t)$. The same information can be extracted using other forcing functions, such as a step change, through the convolution theorems or Laplace transform analysis of linear differential equations.

There are experimental problems with this type of approach that go beyond the obvious ones. The order of the system cannot be determined without information at high frequency ($\omega \gg \sqrt{k/m}$ for our system), and such data are particularly difficult to obtain. The dynamical response therefore cannot be predicted with confidence for times that are smaller than the reciprocal of the maximum frequency for which data are available, regardless of the quality of the data up to that frequency. Other drawbacks are discussed subsequently.

1.2.3 Analogy

Consider the *RLC* (resistance-inductance-capacitance) circuit shown in Fig. 1.3. The equation describing the charge q stored by the capacitance follows

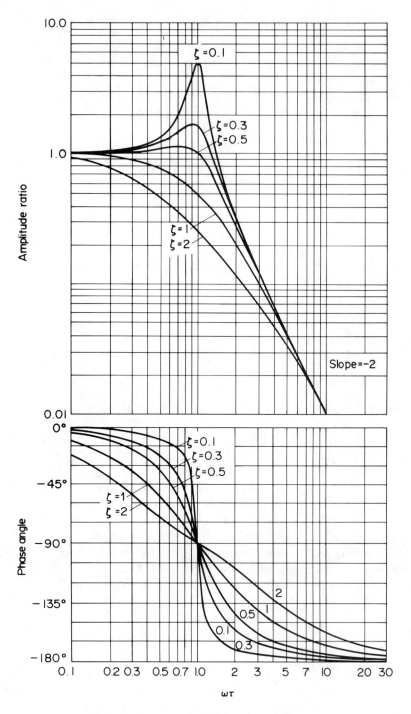

FIG. 1.2.
Bode diagram for second-order system.

1.3 RESERVATIONS AND LIMITS

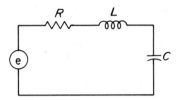

FIG. 1.3.
RLC circuit analogous to mass-spring-dashpot system.

from Kirchhoff's laws of circuits:

$$L\ddot{q} + R\dot{q} + \frac{1}{C}q = e(t) \tag{1.7}$$

e is the imposed voltage, L the inductance, R the resistance, and C the capacitance. This equation assumes linear elements; e.g., Ohm's law applies to the resistor in the circuit. Equation (1.7) is more commonly written in terms of the current, $i = \dot{q}$.

The capacitance is a device for storage of potential energy, similar to a stretched spring, and the charge plays the role of the displacement of the spring from equilibrium. In a similar way, the resistance, like the dashpot, acts to resist and degrade the flow of charge, increasing the effect with increasing current (velocity). The inductance provides inertia, slowing changes in charge effected by the forcing voltage. The analogy is sufficiently suggestive (for we have chosen the words to make it so) to lead us to believe that the displacement of a mass-spring-dashpot system may be described by the equation of an *RLC* circuit. This is the essence of modeling by analogy: the identification of a well-understood process that seems to have the essential features of the process of interest. Analogy to electrical systems is so widespread that it forms the basis of the electrical analog computer, in which complex systems of linear and nearly linear differential equations are solved by constructing the equivalent electrical circuits and then monitoring the time response of the appropriate voltages.

1.3 RESERVATIONS AND LIMITS

The example of modeling by analogy used here is intended to be illustrative, but it is surely not persuasive regarding the potential of this approach, which requires substantial physical intuition. The use of analogy is widespread;

indeed, it is so widespread that it is almost obligatory when writing about the subject to quote for balance the disparaging (and unfair) remark of the physicist and philosopher of science Duhem regarding the contrast between British and continental physics in the early part of the twentieth century.

> The same [i.e., abstraction and generality] does not hold for an Englishman. These abstract notions . . . do not satisfy his need to imagine concrete, material, visible, and tangible things . . . It is to satisfy this need that he goes and creates a model . . . The employment of similar mechanical models, recalling by certain more or less rough analogies the particular features of the theory being expounded, is a regular feature of the English treatises on physics. Here is a book intended to expand the modern themes of electricity and to expound a new theory. In it there are nothing but strings that move around pulleys, which roll around drums, which go through pearl beads, which carry weights; and tubes which pump water while others swell and contract; toothed wheels which are geared to one another and engage hooks. We thought we were entering the tranquil and neatly ordered abode of reason, but we find ourselves in a factory.*

Strings and pulleys will not appear anywhere here, but we will make use of some important relations that have been obtained by analogy. A presumed analogy between the force-displacement relations of various configurations of mass-spring-dashpot systems and the stress-strain relations in macromolecular (polymeric) liquids and solids has been extensively exploited, leading to "Maxwell models" of polymeric liquids and "Kelvin-Voigt models" of polymeric solids, among others. This approach has fallen into disfavor (perhaps because some scientists seemed to take the analogy too literally), and "springs and dashpots" often receives the same contempt as Duhem's strings and pulleys. Nevertheless, the stress equations that we shall use subsequently in describing synthetic fiber manufacture, as well as most other stress formulations in current use, can be traced originally to this analogy.

Similarly, the early kinetic theory of gases, enabling computation of pressure-volume-temperature relations, viscosities, etc., exploited a per-

---

*This is but a brief selection from a sometimes delightful, sometimes maddening essay entitled "Abstract Theories and Mechanical Models" that includes discourses on the mind of Napoleon and the contrast between Corneille and Shakespeare. Only twenty-seven pages after the section quoted above does he note and expand on the thought "The history of physics shows us that the search for analogies between two distinct categories of phenomena has perhaps been the surest and most fruitful method of all the procedures put in play in the construction of physical theories."

ceived analogy between molecular interactions and the collisions of elastic spheres; the latter are described by straightforward Newtonian mechanics. (One often notes such phrases as "The molecules are modeled as hard spheres.") Prandtl's treatment of mass and heat transfer in turbulent flow, which describes these complex phenomena in terms of a single coefficient, is based on a presumed analogy between the motion of turbulent fluid eddies and molecules in a dilute gas; a "mixing length" over which an eddy retains its identity is seen as analogous to the mean free path (average distance between collisions) of gas molecules. The analogy is not really a very good one, but the resulting model has been very useful.

We will not deal with modeling by analogy, as it is too specialized and intuitive. Nor will we deal with empirical modeling (save one brief illustrative example in the wastewater process), although empirical models probably represent the major fraction of models in current use in many fields, for this is the proper realm of statistics and that branch of control theory concerned with stochastic processes. A basic problem with such "black box" models is that they are only interpolations of past data, and can therefore only be used for conditions that are bounded by those for which the data were obtained. Extrapolation beyond these confines, to explore new operating conditions, is dangerous. One of the purposes of modeling (as we shall see) is to search out the unexpected, and an empirical model cannot be trusted when it predicts unexpected phenomena.

Our focus will be on fundamental models, with particular attention to the logical structure of model development and simplification. Fundamental models provide the only means of extrapolation in order to search out new regimes and policies, and hence they should be employed whenever possible. We shall find elements of empiricism and analogy in even the most fundamental models, and this presence will be a major factor in the process of validation.

1.4 A LOGICAL HIERARCHY

There is a logical hierarchy in the construction of mathematical models that is nicely illustrated by the mass-spring-dashpot example. The basic relation that we take as given is Newton's Second Principle, Eq. (1.1). This is a statement of the principle of conservation of linear momentum, and it has universal validity as one of the underlying principles of physics. The spring and dashpot relations, Eqs. (1.2) and (1.3), are specific; we know from experience that springs and dashpots need not be linear. Relations like these are called *constitutive equations*, for they depend on the particular materials and circumstances of use (i.e., on the constitution of the system, hence the name). The constitutive equations in the model are less fundamental than the conservation equation.

Let us suppose, for example, that the spring is finitely extensible; that is, as a critical extension is approached, the incremental force required for further stretching increases at a rate that is much more rapid than linear. Such a spring might be described by the equation

$$F_s = -\frac{kx}{1 - \epsilon|x|} \tag{1.8}$$

Equation (1.1) remains valid, but the dynamics are now described (assuming still a linear dashpot) by the nonlinear equation

$$m\ddot{x} + \mu\dot{x} + \frac{kx}{1 - \epsilon|x|} = F_e(t) \tag{1.9}$$

For small displacements, defined as $\epsilon|x| \ll 1$, Eq. (1.8) is indistinguishable from "Hooke's Law" (which is not a general law at all, of course), and Eq. (1.4) is recovered for the model equation. The nature of the hierarchical structure is clear in the physical and engineering sciences, where the general principles underlying a field can usually be identified and quantified. It seems to be less clearly recognized and exploited in the social sciences. The process of model validation, in which the model predictions are compared to experiment, is often largely a test of the adequacy of the constitutive equations, since the conservation equations are not in doubt. If the distinction between levels of equations noted here does not exist, then the process of validation is considerably more difficult, and confidence in the predictive ability of such a model must be diminished.

BIBLIOGRAPHICAL NOTES

The concept of *model* has received considerable attention from philosophers and logicians. The literature is admirably surveyed by Aris,

R. Aris, *Mathematical Modelling Techniques* (London: Pitman, 1976).

The following, the last three of which are not included in Aris's survey, are of particular interest:

M. B. Hesse, Models in Physics, *British J. Philosophy Science*, **4** (1953) 198.

M. B. Hesse, *Models and Analogies in Science* (South Bend, IN: University of Notre Dame Press, 1966).

P. Achinstein, *Concepts of Science: a Philosophical Analysis* (Baltimore, MD: Johns Hopkins Press, 1968).

M. Redhead, Models in Physics, *British J. Philosophy Science*, **31** (1980) 145.

N. Cartwright, *How the Laws of Physics Lie* (New York: Oxford University Press, 1983).

The selection quoted from Dunhem is from Chapter 4 of the English translation of the second (1914) edition of his *La Théorie Physique: Son Objet, Son Structure*:

P. Duhem, *The Aim and Structure of Physical Theory*, trans. D. D. Weiner (Princeton: Princeton University Press, 1954).

The treatment of mass-spring-dashpot and electrical systems is covered in any elementary book on physics. Frequency response analysis is covered in all introductory books on control theory.

2

WHY MODEL?

2.1 INTRODUCTION

The reason for obtaining a mathematical model is to enable computation of the expected behavior of the process for a range of inputs and conditions. The particular computations to be done depend on the application. A model of the U.S. economy might be used to predict the effects of taxation policies on the gross national product (GNP), for example. A model of a chemical reactor might be used to design a control system to maintain the process within safe operating limits while maximizing production.

An overview of the modeling process is shown in the logical flow diagram in Fig. 2.1. The left-hand vertical column is nothing more than a picture of the classical "scientific method," in which predictions are made and compared with experiment, with revision until the comparison is satisfactory. This process would be devoid of content without the arrows entering from the right, *for it is impossible to separate the development of the model at any stage from the objectives*. The same system might be described by models of varying complexity and quite different structures.

This point is easily demonstrated in a somewhat superficial manner with the mass-spring-dashpot system, which we will assume has a finitely-extensible spring described by the deflection equation given by Eq. (1.8). An "exact" model is then Eq. (1.9):

$$m\ddot{x} + \mu\dot{x} + \frac{kx}{1 - \epsilon|x|} = F_e(t) \tag{2.1}$$

2.1 INTRODUCTION

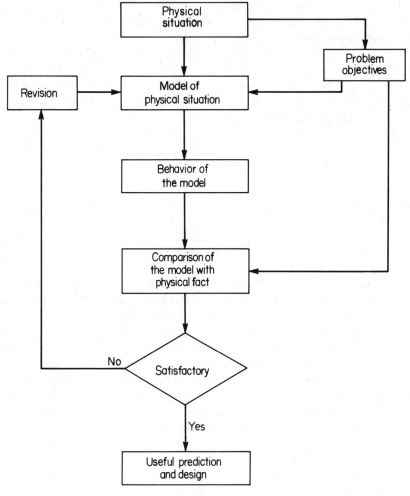

FIG. 2.1.
*Logical flow diagram of the modeling process
(after Russell and Denn, 1972).*

This equation is easily solved numerically for given $F_e(t)$ and initial values of x and $\dot{x}$, but it has no analytical solution. It may be that our interest is only in small forces that cause small displacements about the equilibrium, $x = 0$, so that we will always have $|x| \ll 1$. In that case, the "exact" equation is closely approximated by Eq. (1.4):

$$m\ddot{x} + \mu\dot{x} + kx = F_e(t) \qquad (2.2)$$

This equation can be solved analytically for arbitrary forcing functions (but the solution is meaningful only as long as $|F_e|$ is sufficiently small to ensure that $|x| \ll 1$ at all times); the essential dynamical information is contained graphically in the Bode diagram, Fig. 1.2.

In other situations we might be interested in the displacement caused by forces of arbitrary magnitude, but we might be dealing only with forces that are slowly changing in time.* In that case the terms involving rates of change in Eq. (2.1) will be much less important than the remaining terms, and the resulting model equation becomes

$$\frac{kx}{1 - \epsilon|x|} = F_e(t) \qquad (2.3)$$

or equivalently,

$$x(t) = \frac{F_e(t)}{k + \epsilon|F_e(t)|} \qquad (2.4)$$

Equations (2.2) and (2.4) describe the same process, yet one is a linear differential equation and one is a nonlinear algebraic equation. The former is the appropriate model if our objective entails understanding process behavior for small forces and displacements, but over arbitrary time scales. The latter is appropriate when our objective entails process response for forces of arbitrary magnitude, but slow rates of change.

Further reflection indicates that even the "exact" Eq. (2.1) might be inadequate for some objectives. The dashpot (and the resistor, in the electrical analog) resists motion by degrading energy, leading to a rise in temperature. We might think of a cyclic experiment taking place over many periods. We expect initially to see some heating in the dashpot, causing a decrease in the fluid viscosity and hence in the value of μ. Eventually, it is likely that a steady state will be reached between the dissipation rate and the rate of heat loss to the surroundings, so the temperature (and hence μ) will approach a constant value. We will, therefore, require an equation describing the change in temperature of the dashpot fluid, as well as a relation between this temperature and the dashpot resistance coefficient, μ. The equation describing the temperature change will require use of the principle of conservation of energy, which until now has not even entered our consideration for this system. The coupled energy and momentum equations will then have to be solved simultaneously.

---

*"Slowly changing" is a concept that must be quantified. This is done by proper scaling of the variables, which we discuss in Chapter 6. Roughly, we are restricted here to changes in force that take place over times that are long relative to two time scales: μ/k and m/μ. The former will usually be the more important of the two.

The preceding remarks are intended to emphasize the fact that there is no ONE MODEL for a given process (except the process itself). Rather, there will be many models—indeed, perhaps a continuum of models—each appropriate to given objectives. The development and testing of the model must be done within the framework of these objectives; when a selection from among more than one competitive models is to be made, the principle of parsimony requires that the simplest model consistent with the available observations and objectives be chosen. We shall have occasion in Chapter 13 to examine the concept of "simplest," which appears intuitively to be rather elementary. The concept is in fact quite subtle; consider the challenge of *quantifying* "simplest." The subtleties are not simply academic, but have a real bearing on practical aspects of model selection.

2.2 SETTING THE STAGE

The remainder of this chapter is given over to the descriptions of some processes that will be used subsequently as examples as we study modeling technique. It will be helpful to keep these potential applications in mind, together with the range of objectives associated with them, while reading the general discussions that follow. These applications have been selected because of our personal involvement with each. It is probable that others would choose different, and perhaps better, illustrative systems; the aspect of familiarity outweighs other factors for us, and taken as a whole the examples illustrate the points about modeling that we have found to be important.

2.3 FLUID CATALYTIC CRACKER (FCC)

Catalytic cracking is one of the major steps in the manufacture of gasoline. A high molecular weight fraction of the crude oil known as gas oil, which is distilled from the crude at high temperature, is "cracked" by contact with a catalyst to form low molecular weight hydrocarbon compounds, including gasoline. (Gasoline is a mixture of hydrocarbons containing at least five carbon atoms and boiling up to about 210°C.) Carbon in the form of coke is a product of the cracking reaction. The coke deposits on the catalyst and blocks access to the *active sites*, so the efficiency of conversion decreases with time and the catalyst must be *regenerated* by burning off the carbon. The cracking reactions are *endothermic;* that is, heat must be added in order to keep the temperature from falling. This forms the basis for a highly en-

2 WHY MODEL?

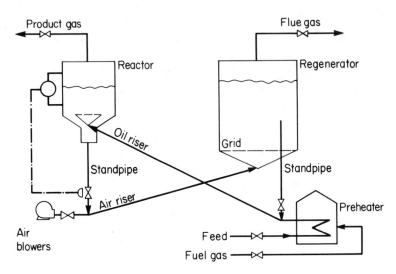

FIG. 2.2.
Schematic of the fluid catalytic cracking process.

ergy-integrated process, as shown in Fig. 2.2. Carbon is burned from the catalyst with air in the regenerator. This heats the catalyst to over 600°C. The hot catalyst is then recycled with the feed gas oil to the cracking reactor, which takes heat from the hot regenerated catalyst to drive the cracking reaction at about 500°C.

Both reactors are fluidized beds. The catalyst particles are very small, typically of order 100 μ in diameter. The reactor is operated in such a way that the pressure drop across the bed of catalyst exceeds the weight of the bed per unit area; thus, the bed is lifted and the catalyst is in a "weightless" state. The velocity is less than that required to entrain the particles and carry them from the bed, so they remain in a fluidized, liquidlike state, moving about randomly with frequent collisions.* The frequent collisions ensure good mixing and efficient heat transfer (particle-particle heat transfer is much more

---

*The gas flow rates to effect fluidization and to entrain and carry over particles are given approximately by the following two equations, respectively:

$$\text{fluidization: flow rate} \geq 9 \times 10^{-4} A(\rho_p - \rho)gD_p^2/\eta$$

$$\text{entrainment: flow rate} \geq 6 \times 10^{-2} A(\rho_p - \rho)gD_p^2/\eta$$

ρ_p and ρ are the densities of particle and gas, respectively; A is the reactor cross-sectional area; g is the acceleration of gravity; D_p is a weighted mean particle diameter; and η is the gas viscosity.

effective than particle-gas-particle heat transfer), so the reactor temperatures remain spatially uniform.

Fluid catalytic cracker models are in common use in the petroleum industry for several purposes. The region of economically attractive operation will change both with changing feedstocks (resulting, perhaps, from different sources of crude oil) and changing product distribution requirements (more heating oil relative to gasoline in winter, for example). Models are used to determine the best operating conditions, which will usually lie against one or more process-imposed constraints. Reactor temperatures and concentrations must be kept within safety and materials limits; too much unreacted oxygen in the regenerator flue gas can cause postcombustion of carbon monoxide and unacceptably high temperatures in the cyclone separator that is used to remove particulate fines from the gas. Models have been used for control system synthesis in order to ensure acceptable dynamical variations of temperatures and conversions in the face of external upsets; the dynamic models have been particularly successful in identifying the source of a sluggish transient response that is not intuititvely apparent from consideration only of the time scales of response of the two reactors.

2.4 MOVING BED COAL GASIFIER

The conversion of coal to clean gaseous products is carried out in a variety of existing and proposed reactor configurations. The objective is to make carbon monoxide, hydrogen, and methane from the solid and volatile hydrocarbons contained in the coal. CO and H_2 are the desired products, in varying ratios, if the gas is to be used for chemical feedstock or power generation. Methane (CH_4) is the desired final product for pipeline substitute natural gas; the gasifier product gas always contains CO and H_2, and additional downstream processing is required for complete conversion to methane.

A schematic of the Lurgi pressurized dry ash moving bed reactor is shown in Fig. 2.3. This reactor configuration has been in commercial use since 1936, and installations exist throughout the world. Coal particles of approximately 10 mm diameter are fed through a lock hopper to the top of the three-to-five meter diameter reactor, which operates at a pressure of about 25 atm. The coal bed moves down through the three-meter reaction zone under the force of gravity, and countercurrent to the gas stream. A rotating grate at the bottom removes ash and serves as a distributor for feed steam and oxygen (or air). The reactor is cooled by a wall water jacket, and 10% to 20% of the process steam is produced by heat transfer from the reactor to the jacket. A typical product stream, obtained in experiments with oxygen

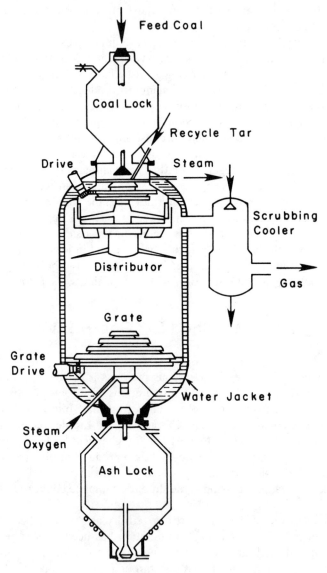

FIG. 2.3.
Schematic of Lurgi gasifier.

gasification of Illinois No. 6 coal, an eastern U.S. bituminous coal, is shown in Table 2.1.

Coal is a complex material. After drying, it contains inorganic matter known collectively as ash; volatiles, which are tars, oil and gases that are

2.4 MOVING BED COAL GASIFIER

TABLE 2.1.
Product composition from gasification of Illinois No. 6 coal in a Lurgi gasifier at Westfield, Scotland.

	Volume % of Dry Gas
H_2	39.1
CO	17.3
CO_2	31.2
CH_4	9.4
C_2H_6	0.7
H_2S + COS	1.1
Inert	1.2

TABLE 2.2.
Typical proximate analysis of Illinois No. 6 coal.

	Weight %
Ash	9.6
Moisture	4.2
Volatile matter	34.2
Fixed carbon	52.0

driven off by gentle heating (the A.S.T.M. test uses heating at 950°C for 7 min in a covered crucible); and the remaining coke residue, which is known as fixed carbon. The latter contains mostly carbon, but some H_2, O_2, N_2, and S as well. The *proximate analysis* of Illinois No. 6 coal is shown in Table 2.2. (This might better be called an Aristotelian analysis.) The *ultimate*, or *elemental analysis* for this coal (on a dry ash free, or daf basis) is shown in Table 2.3. A typical volatile composition obtained by averaging reported values for many coals is shown in Table 2.4.

TABLE 2.3.
Typical ultimate analysis of Illinois No. 6 coal.

	Weight %, daf
C	77.3
H	5.9
N	1.4
S	4.3
O	11.1

TABLE 2.4.
Typical devolatilization data.

	Weight %
Tar, oil plus phenol	20
Chemical water	23
Coal gas	57
Breakdown of coal gas	Volume %
CH_4	50
H_2	13
CO	21
CO_2	6
Others (hydrocarbons, H_2S, N_2)	10

Several important chemical reactions between the solid carbon and gases are readily identified:

$$C + O_2 \rightleftarrows CO_2$$
$$2C + O_2 \rightleftarrows 2CO$$
$$C + CO_2 \rightleftarrows 2CO$$
$$C + H_2O \rightleftarrows H_2 + CO$$
$$C + 2H_2 \rightleftarrows CH_4$$

In addition, a gas-phase reaction* known as the water-gas shift, is very fast and is sure to occur:

$$CO + H_2O \rightleftarrows CO_2 + H_2$$

Other reactions undoubtedly occur as well, but these appear to be the most important in the particular reactor configuration under consideration. The reactions of carbon with oxygen are highly exothermic, and heat must be removed. The gasification reactions of carbon with CO_2 and steam are endothermic, and require heat addition. Direct hydrogenation to form methane is exothermic, but this reaction is probably not very important. The water-gas shift reaction is slightly exothermic.

The exothermic-endothermic reaction system is the basis of the reactor design, which maintains the thermal balance in much the same way as the FCC, but in a single reactor. Consider the schematic in Fig. 2.4 of the events taking place in the reactor, following the coal as it travels through the reactor. First, hot product gas heats and dries the coal. Next, contact with the hot gas drives off the volatiles. The remaining char then comes in contact with a hot gas containing steam and CO_2, but no oxygen; the endothermic gasification reactions therefore take place. Finally, the hot char comes in contact with an oxygen-rich gas, and combustion occurs.

It is helpful to repeat the trip through the gasifier, following the gas flow. First, combustion of char occurs. This uses up all of the oxygen, and drives the temperature up. Since combustion is a very rapid process once ignition has occurred, the temperature rise should be very sharp. Next, the hot, oxygen-free gas, containing feed steam and CO_2 from the combustion reaction, reacts endothermically with char; this drives the temperature down, but at a slower rate than the prior rise, because the gasification reactions are much slower than combustion, and the rate decreases as the temperature drops. There will then be a plateau region as devolatilization occurs, since

---

*The water-gas shift reaction is known to be catalyzed by components of the solid phase, and the speed with which the reaction occurs depends on the nature of the substrate. Thus, the reaction is not really in the gas phase, but at the solid surface.

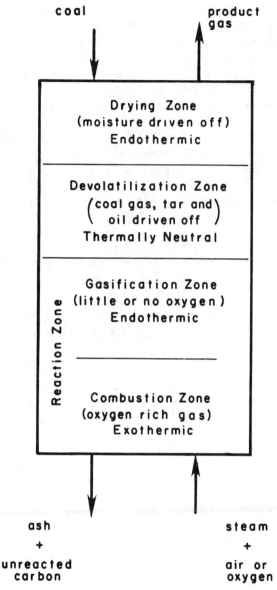

FIG. 2.4.
Schematic of physical and chemical processes in a countercurrent moving bed coal gasifier.

this process seems to be approximately thermally neutral. Finally, the temperature will drop in the drying region. It is interesting to note that this sequence results in volatile evolution above the combustion zone. Thus, while some thermal cracking of the volatiles undoubtedly takes place, combustion does not, and the full heating value of the volatiles is retained in the product gas. Most of the methane in the product gas in Table 2.1 can be accounted for by volatile evolution.

Temperature profiles reported by Lurgi for two coals are shown in Fig. 2.5. (Note that there is no scale on the vertical axis.) The qualitative features of these profiles are consistent with the above description. It is unlikely, however, that these are reliable data. The measurement problem of obtaining meaningful temperature data in the interior of a pressurized jacketed reactor containing over 50 tonnes of moving coal in a corrosive environment seems insurmountable. The "data" in Fig. 2.5 most likely represent indirect inferences, rather than actual measurements.

The most important operating constraint is the maximum tempera-

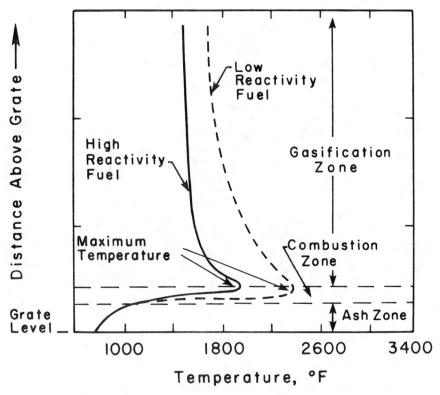

FIG. 2.5.
Reported temperature profiles for high- and low-reactivity coals in a Lurgi gasifier (after Hebden, 1975).

ture. The ash in most coals becomes soft and sticky around 1100 to 1200°C. If the reactor temperature gets this high, then particles will begin to agglomerate and form "clinkers," and the grate will not operate properly.* Six to eight times the steam required for reaction is fed as a temperature moderator. Since the temperature cannot be measured directly, other measurements that can be made (e.g., effluent gas temperature, grate torque, product gas CO_2, etc.) must be used to infer the temperature. This type of inferential control can be (and has been) one of the objectives of modeling the process. As we shall see, certain measurements are sensitive indicators for some coals and not for others.

"Flowsheeting" requires different information, since the objectives are different. Here, we are trying to determine the performance of an entire plant, and examine options regarding equipment selection, configuration, etc. What is required is the composition and temperature of each stream in the plant, and the way in which these properties change with changing feedstock (coal type, for example), operating variables, etc. If a coal gasifier is used to provide gas to a gas turbine in a power plant, for example, only the flow rate and heating value of the gas are needed to determine the power production. The details of the temperature profile are irrelevant in this case, and it is likely (and in fact true) that a simpler model of the process will suffice.

Finally, use of a coal gasifier in a power cycle will require transient operation on two different time scales. Power systems must respond to small demand changes over a time scale on the order of minutes. During off-peak hours, it is likely that some of the gasifiers providing gas for a power plant will be *banked*, or put on hot standby, to be brought back on line as needed; these transients will be on the order of hours. It may be necessary to devise control strategies to effect desired gas production rates and acceptable temperature profiles. (How do we ensure that the combustion zone will not be extinguished during banking because of heat loss to the water jacket, for example?) Still different information is required here.

The range of needs outlined here, from analysis of rapid transients to flowsheeting, is reminiscent of the extremes in model requirements and type that we sketched out for the mass-spring-dashpot system, as illustrated by Eqs. (2.2) and (2.4). We shall indeed see analogous, though far more complex, models for the gasifier.

---

*An alternative configuration, which is not commercial at the time of writing, is the slagging reactor developed by British Gas with Lurgi. The grate is replaced by a hearth, and steam and oxygen are fed through jets (*tuyères*) from the side. Little excess steam is supplied, and the maximum temperature, which occurs at the hearth, is above the temperature at which ash runs freely as a molten slag. The slag is removed through a tap hole in the hearth.

2.5 FIBER SPINLINE

Synthetic fibers are "spun" from polymers, glass, and even metal and pitch. The process might better be called *continuous drawing*, but spinning is firmly embedded in the nomenclature of the textile fiber industry. Our emphasis will be on the manufacture of synthetic fibers from polymers such as nylon, polyethylene terephthalate, and polypropylene. The goal of the manufacturing process is the production of continuous filaments having uniform diameter and desired physical properties with regard to modulus (equivalent to a spring constant), extension-to-break, dye uptake, etc. Nonuniform filaments, when woven into a fabric, will dye nonuniformly, usually in bands.

The spinning process is shown schematically in Fig. 2.6. Molten polymer is extruded through a small hole in the *spinneret* face into quench air that is below the solidification temperature. The solidified filament is wound around a pair of *Godet* rolls which operate at a linear speed that is much greater than the mean extrusion velocity, resulting in a net area reduction relative to the spinneret; the area reduction ratio is known as the *draw ratio*. Most of the draw takes place in the melt region prior to solidification. A liquid finish is usually applied to the solidified filament prior to the Godets, and the filament usually undergoes further solid state processing for property development.

A single filament, as shown in the schematic, will typically be spun in laboratory facilities, but in pilot plant and commercial operation a spinneret will usually contain many holes, sometimes several hundred. The molten polymer is metered through a melt pump to a spinning pack, which consists of a filter bed to remove gel particles and any other impurities. There is then a small flow readjustment region, followed by the spinneret plate. The extruded filaments are taken up together as a *yarn*. The quench air is blown across the yarn, causing filaments in different locations in the yarn bundle to see different quench temperatures and velocity fields, resulting in possible property variations across the yarn.

Typical processing variables for the manufacture of polyethylene terephthalate[*] (PET) fibers are shown in Table 2.5. Tire cord is manufactured at takeup speeds of order 1000 m/min, while textile fiber is manufactured at about 3500 m/min. (Pilot scale equipment that can go up to 8000 to 10,000 m/min is available.) Single filament laboratory experiments rarely achieve takeup velocities approaching 1000 m/min or draw ratios approach-

---

[*]Polyethylene terephthalate is the most common *polyester* fiber. It is sold under a number of trade names, including Dacron, Terylene, and Fortrel. The solid fiber formed under commercial spinning conditions is almost entirely amorphous in structure, although a fiber with up to 45% crystalline regions can be spun.

2.5 FIBER SPINLINE

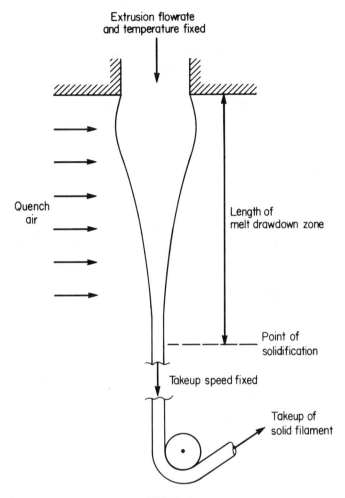

FIG. 2.6.
Schematic of the melt spinning process.

ing 150. Furthermore, laboratory experiments are often carried out for research purposes under isothermal conditions, in which the liquid filament is drawn in an oven that is maintained at the extrusion temperature, followed by rapid solidification at a predetermined point by passage into a cold water bath.

The important solid-state physical properties of a nearly amorphous polymer like PET trend to correlate with the optical birefringence, which is the difference between indices of refraction in two orthogonal directions. Figure 2.7 shows such a correlation with extension-to-break for some pilot

TABLE 2.5.
Typical processing variables for manufacture of PET fibers.

Processing Variable	Typical Value
Extrusion temperature	290° C
Solidification temperature	70° C
Ambient air temperature	35° C
Spinneret hole diameter	0.25 mm
Takeup velocity	1000–3500 m/min
Cross-flow air velocity	0.3 m/sec
Draw ratio	200

scale and commercial plant PET filaments. The primary variables that one wishes to control are therefore birefringence, including radial birefringence variation, and yarn uniformity; the latter is measured (in units of grams/denier!) as mass per unit length. A modern textile fiber plant contains hundreds

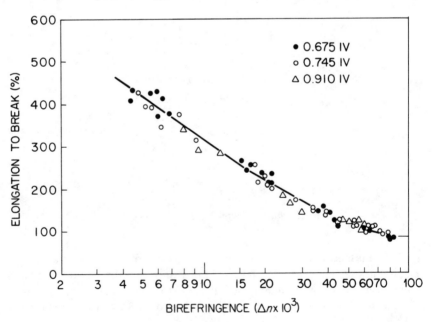

FIG. 2.7.
Correlation between elongation-to-break and birefringence for PET filaments (George, 1982, copyright Society of Plastics Engineers, reproduced with permission).

2.6 WASTEWATER TREATMENT

of spinning positions, so on-line feedback control of birefringence and uniformity is not possible. Filament breakage is perhaps the most serious operating problem, since the machine must be put back on line manually by an operator. The same polymer feed supplies four spinning positions in a modern plant, so loss of one position can sometimes shut down three others.

There has been considerable activity in recent years to develop spinline models in order to study the effects of variables like polymer properties, extrusion temperature, and quench air profile on uniformity, stress level, and birefringence. A computer program developed by the Toyobo Company of Japan that models PET spinning in the steady state is available for commercial licensing at a cost of over $200,000, and at least one U.S. corporation has acquired the rights. Modeling to analyze the propagation of disturbances during transients is less well developed. We shall discuss both aspects. Figure 2.8 is perhaps a tantalizing way to leave this subject for now. It contains data on diameter variation of PET that was extruded and taken up at constant rate, with the drawing taking place at extrusion temperature, followed by rapid solidification in a water bath. This large, regular, self-

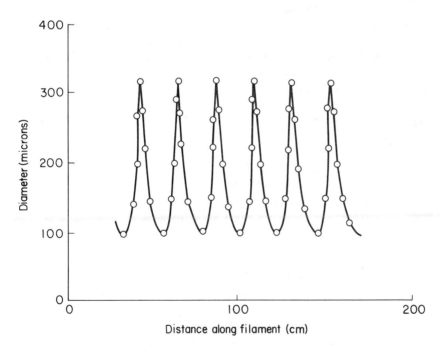

FIG. 2.8.
Diameter variation in a PET filament under steady extrusion and takeup conditions (Ishihara and Kase, 1976, copyright John Wiley and Sons, Inc., reproduced with permission).

sustained thickness variation, known as *draw resonance*, can occur also in the mechanically similar commercial process of extrusion coating of sheets, and can destroy the optical properties of the coating. Spinline modeling has been used successfully to identify the cause of draw resonance and to relate the onset to polymer properties and processing conditions.

2.6 WASTEWATER TREATMENT

The treatment of municipal wastewater is usually carried out by the *activated sludge* process, in which microorganisms metabolize the waste material that would otherwise consume oxygen following discharge into the receiving river, lake, estuary, or sea. Too low an oxygen level will kill fish and other living things, and cause the body of water to "die." The discharge of inadequately treated wastewater by the city of Philadelphia, for example, has resulted in oxygen levels in the Delaware River estuary that are inadequate to sustain migrating fish during the season of low flow.

We will use the activated sludge process as an example of how a model can be used to analyze an entire process, both in terms of design and control. In fact, we will be able to see how controllability can be integrated into the design process through modeling. The process is sufficiently complex that it will also illustrate some limits of the fundamental modeling that we will be developing throughout this book.

The flowsheet is shown in Fig. 2.9. Municipal wastewater is first passed through screens (to remove large objects) and then into a primary settler, where typically about 60% of suspended solids settle out. Some of the suspended solids are carbonaceous and have a *biological oxygen demand* (BOD). The water then passes into one or more reactors, known as aerators, where microorganisms attack the dissolved and suspended carbonaceous material (as well as nitrogen and phosphorous compounds); the microorganisms grow on this substrate, and produce CO_2 and water. Air or oxygen must be added continuously to keep the dissolved oxygen level above a critical value of about two parts per million.

The slurry of treated wastewater and suspended microorganisms then passes into a secondary settler, or clarifier. Clarified, treated wastewater, with BOD and suspended solids levels below the limits established by regulatory agencies, is released to the receiving body of water. The concentrated microorganisms (the activated sludge) are recycled to the reactor, except for a small stream that is removed. This wastage rate must equal the rate at which new biomass is formed in the reactor, or else the microorganism level in the reactor would continue to grow and eventually shut down the process. The sludge that is removed is broken down by anaerobic (no

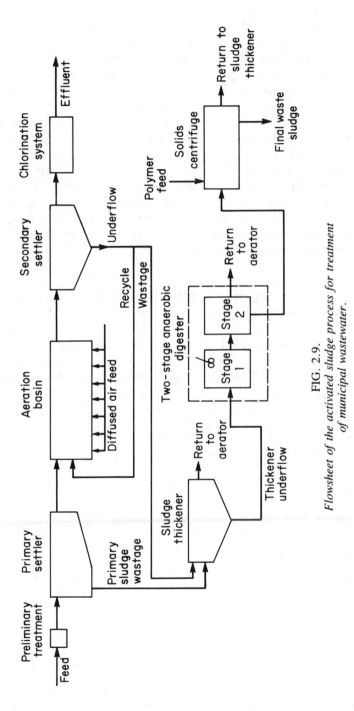

FIG. 2.9.
Flowsheet of the activated sludge process for treatment of municipal wastewater.

oxygen) microorganisms, then dewatered and discarded. (Sludge removal is itself an environmental problem. Some sludge is dumped in the ocean, some is used for landfill. Sludge can be used as a fertilizer, and there is a commercial market, but heavy metals, which can be concentrated by microorganisms, are a potential hazard.)

Our focus in modeling this process will be twofold. First, we will see how a process model, combined with economic information and operating objectives (in this case, environmental constraints on discharged effluent), can be used to analyze design tradeoffs. Second, we will see how rather elementary dynamic modeling that accounts for diurnal variations can be integrated with the results of the design study; this integration enables consideration of process operability to be taken into account in selecting among design alternatives. This full-scale case study will be the subject of a single chapter, but we will occasionally refer to some aspects of the process as we are developing modeling technique.

BIBLIOGRAPHICAL NOTES

Detailed bibliographical references for the processes described here will be given in the chapters where they are discussed in detail: fluid catalytic cracker, Chapter 5; coal gasifier, Chapter 7; fiber spinning, Chapter 12; wastewater treatment, Chapter 14.

Figure 2.1 is based on the discussion in

T. W. F. Russell and M. M. Denn, *Introduction to Chemical Engineering Analysis* (New York, NY: Wiley, 1972).

Figure 2.5 was contained in a paper presented by D. Hebden at the Seventh Synthetic Pipeline Gas Symposium in Chicago, October, 1975. Figure 2.7 is from

H. H. George, *Polymer Engineering and Science* **22** (1982) 292.

Figure 2.8 is extracted from data in

H. Ishihara and S. Kase, *J. Appl. Polymer Science* **20** (1976) 169.

3

CONSTRUCTING A MODEL

3.1   INTRODUCTION

There is a well-defined logical process for constructing a mathematical model. This process essentially formalizes the steps that we followed intuitively in Sections 1.2.1 and 1.3 with the mass-spring-dashpot example. We will describe the logic of model building in this chapter, and then go on in the following chapter to an elementary, but instructive, example. Some definitions will be required. It is important to note that some of the distinctions that we draw in this chapter between hierarchical levels in the logical structure do not seem to be traditionally recognized in the social sciences (nor have they always been recognized in the engineering and natural sciences, with important negative consequences).

3.2   VARIABLES

In most problems of interest, we will wish to be able to describe system response in space and/or time. These are then our *independent variables,* since we are free to specify arbitrary values (within physically acceptable bounds) and expect that the model will contain a computational algorithm that will provide the required outputs. It is helpful to maintain a distinction between these four possible independent variables (time and three spatial coordinates) and model *inputs,* such as the external force in the mass-spring-dashpot system; the latter may sometimes be ours to choose in an arbitrary manner, but they play a very different role in the analysis of the model response.

Modeling of physical systems will always require the application of one or more of the fundamental conservation principles: conservation of mass, momentum, and energy. Specification of these quantities at any time and position contains all of the information that we may require to study any aspect of process behavior. We call these quantities, and their equivalents in nonphysical processes, the *fundamental dependent variables*. Newton's Second Principle is a statement of the principle of conservation of momentum, for example, so the fundamental dependent variable in the mass-spring-dashpot system is momentum.

We have already noted that a model must be expressed in terms of variables that are measurable, at least in principle. This is rarely the case with the fundamental variables. There are no "energy meters." Rather, we measure temperature, pressure, composition, velocity, etc., and from these we infer the energy. Similarly, we compute the momentum from measured velocities and masses. Indeed, even mass is rarely measured directly, but is instead computed from densities and concentrations over known volumes. It is this larger collection of measurable variables that characterizes the fundamental quantities, and the model equations will be written in terms of these *characterizing dependent variables*. It is necessary to select the minimum set of characterizing variables that uniquely defines the fundamental variables. This set defines the *state* of the system.* The characterizing variables for the mass-spring-dashpot system are the velocity and displacement of the center of mass, while time is the sole independent variable.

A first sketch of a logical flow diagram for modeling is shown in Fig. 3.1. Some reflection on our previous example, and the elementary example that follows, will show that this is indeed a start, but a number of important elements are missing.

3.3 A WELL-STIRRED TANK

Consider the process shown schematically in Fig. 3.2. A nonreacting liquid stream consisting of a solvent (say, water) and a dissolved solute (a salt,

---

*There is no generally accepted terminology for what we have called characterizing variables. They are often called *state variables* in the control and systems engineering literature. While this may appear to be an appropriate designation, the term "state variable" has an entirely different meaning in the thermodynamics literature, and thermodynamics is an essential component of the modeling of any process for which energy is a fundamental variable. Thus, we believe that the terminology employed here is less likely to cause confusion.

3.3 A WELL-STIRRED TANK

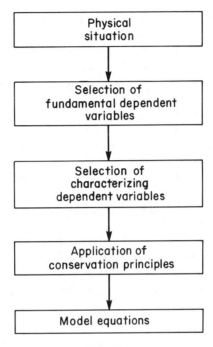

FIG. 3.1.
*Model development flow diagram
(after Russell and Denn, 1972).*

perhaps) flows continuously into a tank containing a solution of the same material, perhaps at a different concentration. An effluent stream is withdrawn continuously. The volumetric flow rate of the feed is denoted q_f, and

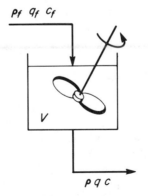

FIG. 3.2.
Well-stirred tank.

that of the effluent is denoted q; these are presumed known as functions of time, as is the concentration c_f of the solute in the feed stream. q_f and q might be measured in units of cubic meters per second (m^3/s), while c_f might be measured in units of kilograms per cubic meter (kg/m^3). It is desired to compute the effluent composition and the liquid volume holdup in the tank as functions of time.

The first thing that we need to do is to select the fundamental conservation principles that are to be applied, and hence the fundamental dependent variables. Conservation of mass is clearly relevant. We will presume that the system is not heated or cooled, and that there are no thermal effects on mixing (such as occur when sulfuric acid streams of different composition are mixed together). We will also presume that we are operating at moderate pressure, so that compressibility of the liquid is not a factor. In that case, the conservation of energy is of secondary importance. We will presume that there is vigorous agitation in the tank, so the principle of conservation of momentum would be expected to be quite important. As we shall see, however, explicit consideration of momentum conservation is often (indeed, usually) circumvented by making appropriate *a priori* assumptions about the flow patterns.

The principle of conservation of mass is simply a bookkeeping statement:

> *The rate of change of mass within a specified region of space equals the rate at which mass enters that region, minus the rate at which mass leaves.*

The designated spatial region is known as a *control volume,* and it is separated from the surroundings by a (possibly imaginary) *control surface.* Selection of the control volume is clearly an important factor in modeling, since we cannot otherwise apply the conservation principle. The selection of the control volume is intimately related to our perception of momentum transport within the system. The simplest assumption that can be made here is that agitation is sufficient to make the system completely homogeneous. In that case, any characterizing variable that is measured at one point in the tank at a given time will have the same value if measured elsewhere in the tank at the same time. This *perfect mixing assumption* therefore allows us to take the tank itself as the control volume; it simultaneously removes any need to consider conservation of momentum in modeling the process. We shall return to this important point subsequently.

The mass is characterized by the liquid density, which we denote ρ; the solute concentration, c; and the liquid volume, V. Total mass is density times volume, ρV, while solute mass is solute concentration times volume, cV. Because of the perfect mixing assumption, the density and concentration

3.3 A WELL-STIRRED TANK

at the effluent have the same values as they have at all points in the tank, so we do not require a separate designation (such as ρ_e or c_e). Mass enters at a rate $\rho_f q_f$, where ρ_f is the density of the feed stream, and leaves at a rate ρq. (Density has units of, say, kilograms per cubic meter, kg/m^3, so ρq has units (kg/m^3) × (m^3/s) = kg/s.) The principle of conservation of mass is, therefore,

$$\frac{d(\rho V)}{dt} = \rho_f q_f - \rho q \qquad (3.1)$$

rate of change of   rate at   rate at
mass contained in   =   which mass − which mass
the control volume   enters   leaves

Conservation of mass also applies to each component species. The total amount of solute in the tank is cV, while the rates at which solute enters and leaves the control volume are, respectively, $c_f q_f$ and cq. Hence,

$$\frac{d(cV)}{dt} = c_f q_f - cq \qquad (3.2)$$

We could write the equation describing the application of mass conservation to the solvent, but that equation would simply be the result of subtracting Eq. (3.2) from (3.1).* The total number of linearly independent equations that can be obtained from conservation of mass is equal to the number of component species in the system.

We now come to another critical aspect of modeling that was alluded to earlier in the discussion of the mass-spring-dashpot system. These two equations exhaust the independent information that is available in the conservation principles. Yet, given that the properties of the feed stream are known at all times, there are three unknowns: V, ρ, and c. It is clear that we lack sufficient equations to solve for the unknowns. This particular example has been chosen because the problem is nicely illustrated, but the resolution is transparent: there must be a unique relationship between the density and the concentration,

$$\rho = \rho(c) \qquad (3.3)$$

This relationship is not derivable from the conservation equations at any

---

*The concentration of solvent in mass units is $\rho - c$, since ρ equals the total mass and c the mass of solute in a unit volume. Thus, the rate of change of mass of solute is $d[(\rho - c)V]/dt$, while rates of entering are $(\rho_f - c_f)q_f$ and $(\rho - c)q$, respectively.

macroscopic level, and it is unique to the particular solute-solvent system being processed. Such a relationship between the characterizing variables that is unique to the constitution of the particular system is called a *constitutive equation*. (In thermodynamics, such an equation is called an equation of state.) Selection of constitutive equations is one of the most important aspects of process modeling, and it is here that judicious experimentation is often required.

Equations (3.1) through (3.3) form the complete dynamical model. If the volume and concentration are known at any time, then ρ, V, and c can be found for all times for arbitrarily specified flow rates and feed composition. It is useful to continue this illustrative example a bit further, however, for well-mixed tanks play an important role in process analysis. Let us suppose that Eq. (3.3) has the particular form

$$\rho = \rho_0 + \rho_1 c + \phi(c) \tag{3.4}$$

where ρ_0 and ρ_1 are constants and $\phi(c)$ contains any nonlinearities in the function $\rho(c)$. It is now a straightforward manipulation to eliminate the density from Eq. (3.1). Substitution of Eq. (3.4) into (3.1), followed by subtraction of Eq. (3.2) multiplied by ρ_0, leads to

$$\frac{dV}{dt} = \frac{\rho_0 + \phi(c) - c_f \phi'(c)}{\rho_0 + \phi(c) - c\phi'(c)} q_f - q \tag{3.5}$$

The prime denotes differentiation with respect to c. If the density function is linear over the interval from c_f to c, then $\phi'(c)$ is zero and Eq. (3.5) reduces to the simpler form

$$\frac{dV}{dt} = q_f - q \tag{3.6}$$

Equation (3.6) could have been obtained from Eq. (3.1) by making the assumption that $\rho_f = \rho =$ constant. This assumption is nearly always made, and is justified on the basis that "all liquid densities are about the same." (It is so commonly done that many practicing engineers believe that volume must be conserved in a liquid system!) In fact, the normal range of liquid densities is from about 800 to 1200 kg/m^3 (specific gravities from 0.8 to 1.2), which is hardly constant; the latter is the value for an aqueous sodium chloride solution with a salt concentration of about 320 kg/m^3. It is the approximate linearity of the density function, apparently corresponding to a *thermodynamically ideal solution*, that justifies use of Eq. (3.6) for most liquid systems. The coefficient of q_f in Eq. (3.5) will not differ substantially from unity even for nonzero $\phi(c)$. (The treatment in this section is a bit superficial, and the situation is more subtle, as is so often the case when thermodynamic considerations enter. We return to the issue of the density-concentration constitutive equation at the end of this chapter.)

3.4 MODELING LOGIC

We should now return to the logical flow sheet in Fig. 3.1. Two essential elements are clearly missing. First, the selection of the control volume, which is of major importance, must be added. This is done in the revised flow sheet in Fig. 3.3. We will return to a brief discussion of the control volume in the next section.

Second, we need to include the constitutive equations in the logical structure. When the conservation principles have first been applied, it is most unlikely that we will have a sufficient number of equations to solve for all of the characterizing variables;* as we saw in the preceding section, this is the case with even the simplest fluid mixing problem. It may sometimes be true that one of the fundamental variables has been inadvertently neglected, and that including it will provide the missing equation—energy conservation to provide an equation for temperature, for example. This possibility should always be considered, and is included in Fig. 3.3 under the query "Conservation equations fully exploited?" It is more probable that the missing equations reflect the need for constitutive equations that describe relations between the characterizing variables.

3.5 CONSTITUTIVE EQUATIONS

The loop introducing the constitutive relations in Fig. 3.3 is in a sense an operational definition: *constitutive equations are those additional relations between characterizing variables that are required for a complete mathematical description.* Constitutive equations are usually associated with molecular phenomena. We have seen a few examples already: the relation between density and concentration in a mixture, and the modulus or spring constant of a stretched elastic material. One of the constitutive equations

---

*We shall generally use the convenient and common shorthand "as many equations as unknowns." This is not a rigorous equivalent, as the counterexample of finding real solutions x and y to the single equation $x^2 + y^2 = 0$ illustrates.

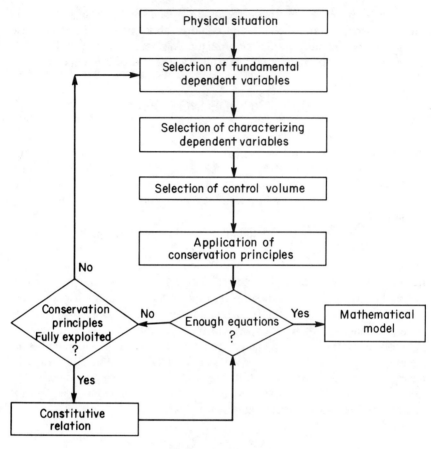

FIG. 3.3.
*Model development flow diagram
(after Russell and Denn, 1972).*

most familiar to physical scientists is the "ideal gas law,"

$$M_w p = \rho RT \tag{3.7}$$

Here, p is the pressure, ρ the mass density, T the absolute temperature, M_w the average molecular weight, and R a dimensional constant known as the gas constant; $R = 8314.3$ m³Pa/kg mol °K.

The ideal gas law illustrates some important features of many constitutive equations. First, it is *not* a law of nature at all, if we take the word "law" to mean "universally applicable." It is an experimentally observed relation that works well for gases at pressures and temperatures that are far

3.5 CONSTITUTIVE EQUATIONS

removed from the critical region, and it is totally inadequate for dense gases. Many other "laws" (Fick's law of diffusion, Newton's law of cooling, Fourier's law of conduction, etc.) are similarly empiricisms that have been given a name implying universality.* Like many constitutive equations, the ideal gas equation can be derived from the conservation principles by making specific assumptions about molecular interactions. This requires an idealization of the molecule, and often the use of a mechanical analogue—gas molecules modeled as elastic spheres in this case, polymer molecules in solution modeled as beads connected by springs or freely jointed rods, etc. (Note that we are using the word "model" in the sense of Section 1.2.3.) The nineteenth-century natural philosopher would probably argue that our constitutive equations can always be derived from conservation principles, if we apply the conservation principles at a sufficiently microscopic level. This argument fails at any reasonable degree of difficulty, however, because we must always specify the form of intermolecular forces—as well as solve the resulting equations for the molecular dynamics, which could take years for each new situation!

Constitutive equations will thus come in most cases from experiment, usually guided by some theory and perhaps dimensional analysis or other invariance arguments. Dimensional analysis is a powerful tool: The ideal gas equation, for example, follows immediately from dimensional analysis if we hypothesize that the pressure in a gas is a function only of the concentration and of the mean energy, where the latter is measured by the absolute temperature. Many constitutive equations are available in the form of dimensionless engineering correlations; the rate of heat transfer from a solid body to a surrounding fluid depends on physical properties and relative mo-

---

*The danger of this tendency to decree new laws of nature is that the "laws" will be viewed as such by students and inexperienced practitioners. The field of fluid mechanics is an excellent example. The relevant constitutive equation here is the one relating stress to deformation rate. Low molecular weight liquids and all gases tend to follow a linear relationship named for Newton, containing two phenomenological coefficients: the shear viscosity and the bulk viscosity. When taken together with the equation of conservation of momentum, one obtains (for isothermal flow at low pressure, although even this restriction is rarely noted) a set of equations in velocity and pressure known as the *Navier-Stokes* equations. (See Section 6.6). These equations are very specific, because they encompass the Newtonian fluid constitutive equation, and they are totally inapplicable to complex liquid materials like molten polymers. The presentation of the Navier-Stokes equations in classical fluid mechanics texts has rarely distinguished between the conservation equations and this particular constitutive assumption. (From the 1979 edition of a widely used fluid mechanics text: "This chapter deals with real fluids . . . The equations of motion for a real fluid can be developed from consideration of the forces acting on a small element of the fluid, including the shear stresses generated by fluid motion and viscosity. The derivation of these equations, called the *Navier-Stokes* equations, . . .") As a consequence, few engineers trained prior to the nineteen sixties (and perhaps since) recognize that the Navier-Stokes equations are limited to a particular class of fluids, and there have been many incorrect attempts to apply solutions of the Navier-Stokes equations to the flow of polymers, suspensions, and other complex liquids.

tion, for example, and the relation is usually presented in the form of a correlation between a dimensionless heat transfer coefficient known as the Nusselt number and dimensionless groups known as the Reynolds, Grashof, and Prandtl numbers, which involve flow properties, physical parameters, and geometry (Section 6.6.3).

We will see many examples of different types of constitutive equations throughout this work. Some will be algebraic, as in the examples cited thus far. The constitutive equation relating stress to deformation rate that we will use in the modeling of fiber spinning will be a partial differential equation.

3.6 CONTROL VOLUME

There is one important restriction on the selection of a control volume. We must be sure that if we insert a probe (conceptually or in fact), then we will measure the same value at each point in the control volume. The physical size of the control volume is thus seen to depend both on the spatial distribution of the characterizing variables and on the degree of discrimination of the measuring probe (or equivalently, our need for discrimination).

In the example in Section 3.3, we assumed that the tank was well mixed, so the entire tank was an appropriate choice for the control volume. Consider instead the primary settler for the secondary clarifier in the activated sludge process, Fig. 2.11. Here, we intentionally set up a gradient in suspended solids from top to bottom through the action of gravity, so that we can withdraw a concentrated slurry from the bottom and a stream from the top containing a lower solids loading than the feed. A measurement of slurry density or solids concentration taken at the top would differ from one taken at the bottom; thus, if either density or solids concentration in the settling tank is to be used as a characterizing variable, the control volume must be chosen on a scale much smaller than the entire tank. The usual practice when there is likely to be a continuous variation in a characterizing variable is to take a control volume that includes a differential slice of the system in the direction of variation, as in Fig. 3.4. In that case, the control volume is itself spatially dependent, and the resulting equations following application of the conservation equations will include one or more spatial coordinates as independent variables. It is clear that the problem objectives define the spatial discrimination that we seek in the characterizing variables, and hence the choice of control volume.

There is some nomenclature that is commonly used and worth noting. When spatial variation is not of interest then we speak of the system as being *lumped*, or *lumped parameter*. Conversely, a system in which spatial vari-

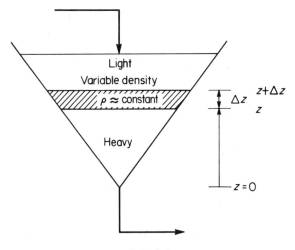

FIG. 3.4.
Differential control volume.

ation is important is called *distributed*, or *distributed parameter*. (Aris notes that the nomenclature "seems misguided, for it is variables, not parameters that are lumped or distributed." The terminology is, however, firmly entrenched.) Distributed systems will generally be described by partial differential equations or integrodifferential equations, while lumped systems will usually be described by ordinary differential equations (or algebraic equations if changes in time are not of interest). The same system might have either description, depending on our modeling objectives and our choice of a control volume. The computational problems associated with distributed systems are usually greater than those associated with lumped systems, and in fact most numerical schemes reduce distributed systems to nearly equivalent lumped systems; we shall discuss this topic to a limited extent in Chapter 10. Conversely, there are some classes of lumped systems that are more easily analyzed in terms of nearly equivalent distributed systems, as discussed in Chapter 11.

3.7 REPRISE: DENSITY OF LIQUID SYSTEMS

The relationship between density and concentration of a liquid system is governed by the intermolecular forces. An ideal mixture of A and B is one in which the forces between molecules of A and A, B and B, and A and B are all the same. In such a case, we would expect that the net volume V

obtained by adding together volume V_A of A and V_B of B would equal $V_A + V_B$, since B molecules in the mixture will react to their new neighbors in the same way as to their old ones. Such will not always be the case, as we know from practical experience; a mixture of equal parts of ethyl alcohol and water at 25°C has a volume that is five percent less than the sum of the two volumes of pure liquid prior to mixing.

The density of a liquid mixture of N species is

$$\rho = c_1 + c_2 + \cdots + c_N \tag{3.7}$$

where $\{c_i\}$ are the concentrations of all species in mass units (e.g., kg/m³). Equation (3.7) is really a definition of the density, and it is used in establishing that the overall conservation-of-mass equation is simply the sum of the component equations. A thermodynamic relation known as the *Gibbs-Duhem equation* establishes that the density of a liquid mixture at constant temperature is a unique function of $N - 1$ concentrations.

The thermodynamics of mixtures is built around *partial molar quantities*. (Recall that a *mole* is equal to Avogadro's number, or 6.02×10^{23}, of molecules. The molecular weight is the mass of one mole.) We shall be concerned here with the partial molar volume. Other partial molar quantities will be introduced in Chapter 5, where we will be concerned with application of the principle of conservation of energy. For convenience, we will work only with binary (two-component) systems.

Consider a large volume V of a liquid, containing n_A moles of A and n_B moles of B; V is uniquely determined by n_A and n_B. At constant temperature and pressure we add a differential amount of A and measure the differential volume change; this experiment defines the *partial molar volume*, $\tilde{V}_A$, as

$$\tilde{V}_A = \left.\frac{\partial V}{\partial n_A}\right)_{T,p,n_B} \tag{3.8a}$$

$\tilde{V}_B$ is defined equivalently as

$$\tilde{V}_B = \left.\frac{\partial V}{\partial n_B}\right)_{T,p,n_A} \tag{3.8b}$$

$\tilde{V}_A$ and $\tilde{V}_B$ are intensive properties, and depend only on the molar *ratio*, n_B/n_A. It is a consequence of the mathematics of exact differential forms, or of an equivalent physical argument, that

$$V = n_A \tilde{V}_A + n_B \tilde{V}_B \tag{3.9}$$

It then follows that

$$n_A \frac{d\tilde{V}_A}{dt} + n_B \frac{d\tilde{V}_B}{dt} = 0 \tag{3.10}$$

Equation (3.10) is a form of the Gibbs-Duhem equation, which establishes the relationship between the partial molar volumes. The partial molar volumes will be constants in an ideal system, with values equal to the molar volume of the pure liquid.

It is convenient for our purposes to define the *partial densities*,

$$\bar{\rho}_A = M_{wA}/\tilde{V}_A, \qquad \bar{\rho}_B = M_{wB}/\tilde{V}_B \tag{3.11}$$

The partial densities will be functions of the mass concentration ratio, c_B/c_A; for an ideal system, $\bar{\rho}_A$ and $\bar{\rho}_B$ will equal the densities of pure A and B, respectively. Now, from Eq. (3.7),

$$\rho = c_A + c_B = \frac{n_A M_{wA}}{V} + \frac{n_B M_{wB}}{V} = \frac{1 + c_B/c_A}{1/\bar{\rho}_A + (1/\bar{\rho}_B)(c_B/c_A)} \tag{3.12}$$

where we have made use of $c_A = M_{wA} n_A/V$ and Eq. (3.9). Equation (3.12) can be rearranged to the form

$$\rho = \bar{\rho}_A + \left(1 - \frac{\bar{\rho}_A}{\bar{\rho}_B}\right) c_B \tag{3.13}$$

Since $\bar{\rho}_A$ and $\bar{\rho}_B$ are functions only of c_B/c_A or, equivalently, c_B/ρ, Eq. (3.13) establishes a unique but implicit relationship between ρ and c_B. This justifies the use of Eq. (3.3) and all that followed from it.

In an ideal solution, $\bar{\rho}_A$ and $\bar{\rho}_B$ are constants, so Eq. (3.13) establishes a linear relationship between density and concentration. The range of ideality will usually be small and restricted to very low concentrations. It may (and often will) be, however, that the implicit relation between ρ and c_B defined by Eq. (3.13) will be linear, with an intercept that differs from ρ_A, over a finite range for c_B different from zero. In that case, the consequences of linearity shown in Section 3.3 still follow, although the solution is not thermodynamically ideal.

BIBLIOGRAPHICAL NOTES

The logical framework laid out here was developed in

T. W. F. Russell and M. M. Denn, *Introduction to Chemical Engineering Analysis* (New York, NY: John Wiley, 1972).

It is also sketched out in

> M. M. Denn, "Modeling for Process Control," in C. Leondes, ed., *Control and Dynamic Systems*, Vol. 15 (New York, NY: Academic Press, 1979).

See also the chapter by Aris entitled "Method in the Modeling of Chemical Engineering Systems," in the same collection, and his monograph

> R. Aris, *Mathematical Modelling Techniques* (London: Pitman, 1978).

The concept of partial molar quantities and derivation of the Gibbs-Duhem equation can be found in any book dealing with solution thermodynamics or physical chemistry. The physical argument is given in Russell and Denn.

4

MODELING REACTORS: CONSERVATION OF MASS

4.1 INTRODUCTION

Many of the important points about modeling are nicely illustrated by the example of a well-stirred tank containing a homogeneous fluid in which a single chemical reaction is taking place. In particular, we can see several ways in which constitutive equations enter and determine system response, often leading to a type of response that may be unexpected. We need to be aware of these possible types of system response, since unexpected and unusual behavior can be an important discriminator of process models. We will look at applications of the principle of conservation of mass in this chapter. We will look at conservation of energy in the next, where we will focus on some problems and common, potentially serious modeling errors.

The reactor is shown schematically in Fig. 4.1. Chemical species A, B, etc. are contained in a continuous liquid feed stream; it may be that each reactant is contained in a separate feed stream, but we will assume for convenience that these streams join and enter the reactor together. The following chemical reaction occurs:*

---

*Some explanation may be necessary for readers who have not worked recently with chemical equations. Equation (4.1) means that α molecules of A combine with β molecules of B (and whatever other reactants, if any, are indicated by . . .) to form μ molecules of M and ν molecules of N, the products. The *stoichiometric coefficients* α, β, μ, ν, . . . will be

4 MODELING REACTORS: CONSERVATION OF MASS

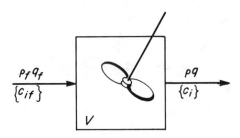

FIG. 4.1.
Schematic of a continuous flow stirred-tank reactor.

$$\alpha A + \beta B + \cdots \rightarrow \mu M + \nu N + \cdots \qquad (4.1)$$

The reactor temperature is controlled by a cooling or heating jacket. Such reactors are in common use industrially, although it is unusual for so simple a reaction scheme to occur. We will develop a model that will enable us to predict reactor temperature and composition as we vary operating conditions.

4.2 CONSERVATION OF MASS

By assuming that the reactor is well stirred, we are able to take the entire tank as the control volume. Mass is characterized by the density, ρ; concentrations (in molar units) c_A, c_B, c_M, c_N, etc. of A, B, M, N, respectively; and liquid volume V. The volumetric flow rate is q. The properties of the feed stream are denoted with a subscript "f"; the effluent is unsubscripted, since it is a consequence of the well-stirred reactor assumption that density and concentrations of the effluent are the same as those in the tank.

The principle of conservation of mass as applied to the total mass in the system is unchanged by the fact of chemical reaction, and is identical to Eq. (3.1):

---

integers, or ratios of integers, except in unusual cases. The reaction could be reversible, in which case A and B would be formed from the products. In an autocatalytic reaction, one or more of the reactants appear in a greater amount as a product (that is, M and A may be the same chemical species, with $\mu > \alpha$). Some examples of chemical equations are given in Section 2.4 in the description of the coal gasification reactor. Concentrations of species that react are usually measured in *moles* per unit volume. A mole of a species contains Avogadro's number (6.02×10^{23}) of molecules; moles are converted to mass by multiplying by the molecular weight, M_w, which has units of mass/mole.

4.2 CONSERVATION OF MASS

$$\frac{d}{dt}\rho V = \rho_f q_f - \rho q \tag{4.2}$$

The equations for conservation of mass of each of the component species must account for the fact that a species can enter or leave the control volume by another mechanism besides flow in and out; each species can also enter or leave the control volume by being created or destroyed in a chemical reaction. Let r_i be the net molar rate of formation of species i by chemical reaction per unit volume;* r_i will be negative if species i disappears in the reaction and positive if species i is formed. The equation of conservation of mass for species A is therefore

$$\frac{d}{dt} c_A V = q_f c_{Af} - q c_A + V r_A \tag{4.3a}$$

Similarly, for any species $(A, B, M, N,$ etc.) the equation of conservation of mass is

$$\frac{d}{dt} c_i V = q_f c_{if} - q c_i + V r_i \tag{4.3b}$$

If the system is operated at constant temperature and pressure, then the conservation of energy is irrelevant and Eqs. (4.2) and (4.3) comprise the full set of relations available from the conservation equations. (Momentum conservation has been made irrelevant by assuming perfect mixing.) The variables include the volume, V; the density, ρ; the concentrations, $\{c_i\}$; and the rates, $\{r_i\}$. The density must be related to the concentrations,† as

---

*In liquid and gas phase reactions, the reaction rate is specified on a unit volume basis, since, all other things being equal, the net rate of formation will increase in direct proportion to the volume. The rate is usually specified on a unit area basis for reactions that take place at a phase interface, such as reactions between a gas and a solid (as in coal gasification), or reactions that are catalyzed at active sites on a solid surface.

†The assumptions necessary for Eq. (4.2) to simplify to $dV/dt = q_f - q$ are somewhat more stringent for a reacting system. Conservation of mass requires

$$\sum_{i=1}^{n} M_{wi} r_i = 0,$$

where n is the total number of species present. Suppose that

$$\rho = \rho_0 + \sum_{i=1}^{n-1} \phi_i c_i.$$

Simplification is then obtained if (1) all ϕ_i are equal to a constant multiplied by the species molecular weight M_{wi} for species for which $r_i \neq 0$; and (2) at least one species, denoted here by n, is inert (a solvent, say), $r_n = 0$. If all densities are really identical, of course, then the simplification is trivial.

before, but it is clear that we must be able to relate the rates to the concentrations as well.

One property of the rates follows immediately from the chemical equation, Eq. (4.1). Whenever α moles of A react, so do β moles of B; thus, $r_A/\alpha = r_B/\beta$. Furthermore, μ moles of M and ν moles of N are formed whenever α moles of A vanish, so $r_M/\mu = r_N/\nu = -r_A/\alpha$, and equivalently for any other species present. Thus, we really need only a single rate, which we denote r. There is no loss of generality in taking $\alpha = 1$, since we can always divide Eq. (4.1) by a positive constant and retain equality. If we then define r as the net rate of disappearance of A by reaction per unit volume, we can write Eqs. (4.3) as

$$\frac{d}{dt} c_A V = q_f c_{Af} - q c_A - Vr \qquad (4.4a)$$

$$\frac{d}{dt} c_B V = q_f c_{Bf} - q c_B - \beta Vr \qquad (4.4b)$$

$$\frac{d}{dt} c_M V = q_f c_{Mf} - q c_M + \mu Vr \qquad (4.4c)$$

r is often referred to as the *intrinsic reaction rate*.

The rate of reaction must depend on the amounts of reactants present; since we are looking at a rate per unit volume, this dependence must in fact be on the concentrations. The particular form of this constitutive equation must always be obtained experimentally, although theory can sometimes lead to the functional form and even occasionally to orders of magnitude of the constants. Reaction rates are usually measured in *batch reactors*, with $q_f = q = 0$. With constant volume the mass balance equations reduce to $dc_A/dt = -r$, so the rate function is obtained directly in a straightforward manner by measuring concentrations as a function of time. The rate equation obtained in this way is, of course, a constitutive equation that describes this reacting system in *any* reactor configuration, not only the one in which the measurements were made. There is a considerable amount of confusion over this seemingly elementary point, particularly in the environmental engineering literature.

We can make some general statements about the form of r. The rate of disappearance of any reactant will usually increase monotonically with the concentration of that reactant, and the rate must vanish if the concentration of the reactant goes to zero. Thus, we will often see rate expressions of the form

$$r = k c_A^a c_B^b \cdots \qquad (4.5)$$

It would be unusual for the rate of disappearance of a reactant to depend on the concentration of a product, although "product inhibition" is sometimes seen, particularly in biological reactions. If the reaction is *reversible*, by which we mean that μ moles of M and ν moles of N can also react to form α moles of A and β moles of B, then the net rate of disappearance of A must also include the rate of formation; we then expect expressions like

$$r = k_1 c_A^a c_B^b \cdots - k_2 c_M^m c_N^n \cdots \tag{4.6}$$

These power functions are not the only constitutive forms possible or even in use, and they are often inadequate, but they are frequently employed because they fit available data over a relatively wide range. The *rate constants* k_1, k_2, etc. are usually temperature dependent. In some cases the exponents in the power functions are the same as the stoichiometric coefficients ($a = \alpha$, $b = \beta$, $m = \mu$, $n = \nu$, etc.); this would follow from an elastic sphere model of molecules, in which the rate of reaction is proportional to the probability of a collision between reactive spheres.

4.3 A DESIGN EXAMPLE

The use of a model for studying design tradeoffs can be illustrated with a system like that shown in Fig. 4.2. The irreversible reaction

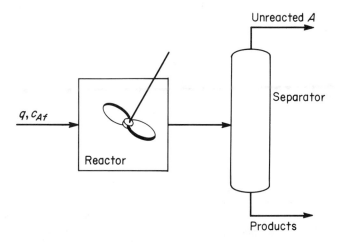

FIG. 4.2.
Schematic of reactor followed by product separation.

4 MODELING REACTORS: CONSERVATION OF MASS

$$A + \beta B \rightarrow \text{products}$$

is to be carried out in a well-stirred reactor, followed by removal of unreacted A for product purity. We wish to design the system for steady-state operation to meet a specified production rate. We will assume that the supply rate of A is fixed, perhaps because it comes from another processing system, but that the supply rate of B may be specified by the designer. For simplicity, we will suppose that the rate constitutive equation is of the most elementary form possible:

$$r = kc_A c_B \qquad (4.7)$$

More complex forms would make the algebra more difficult, but would not change any essential features.

All time derivatives vanish at steady state. If we assume that the density is linear in reactant and product concentrations, and that the reaction is carried out in an inert solvent, then $dV/dt = 0$ implies $q_f = q$. Equations (4.4) for the reactants then become

$$0 = q(c_{Af} - c_A) - Vkc_A c_B \qquad (4.8a)$$

$$0 = q(c_{Bf} - c_B) - V\beta kc_A c_B \qquad (4.8b)$$

The production rate (PR) is proportional to the rate of conversion of A, so we have

$$PR = q(c_{Af} - c_A) = \text{fixed} \qquad (4.9a)$$

The feed rate (FR) of A is also fixed:

$$FR = qc_{Af} = \text{fixed} \qquad (4.9b)$$

Equations (4.8) can then be manipulated to the equivalent forms

$$PR = k(FR - PR)\frac{V}{q}c_B \qquad (4.10a)$$

$$qc_{Bf} = qc_B + \beta PR \qquad (4.10b)$$

These last two equations establish two relations between the four process variables still unspecified: volume, V; flow rate, q; feed rate of reactant B, qc_{Bf}; and effluent concentration of B, c_B. There are thus two degrees of freedom available to the designer.

The process cost will be a sum of capital and operating costs. Op-

4.3 A DESIGN EXAMPLE

erating costs will typically vary with the throughput, q. Reactor capital cost will vary with volume, V. Separation costs will depend on the concentration of the material to be removed and on the throughput. Raw material costs will depend on the feed rate of B. For illustration, we will take all cost terms to be linear (i.e., constant marginal rate of return); more realistic economics would only require more algebraic manipulation. We can then write

$$\text{Cost} = \underset{\substack{\text{reactor}\\\text{capital}}}{\lambda_1 V} + \underset{\substack{\text{operating, and}\\\text{portion of separation}\\\text{dependent on throughput}}}{\lambda_2 q} + \underset{\substack{\text{separation}\\\text{of } A}}{\lambda_3 c_A}$$
$$+ \underset{\substack{\text{separation}\\\text{of } B}}{\lambda_4 c_B} + \underset{\substack{\text{raw}\\\text{material}}}{\lambda_5 q c_{Bf}} \quad \textbf{(4.11)}$$

With Eqs. (4.9) and (4.10), this cost function can be written in terms of two variables, representing the degrees of freedom available to the designer; we choose q and c_B. We then have

$$\text{Cost} = \frac{\lambda_1 \text{PR}}{k(\text{FR} - \text{PR})} \frac{q}{c_B} + \lambda_2 q + \frac{\lambda_3(\text{FR} - \text{PR})}{q}$$
$$+ \lambda_4 c_B + \lambda_5 q c_B + \lambda_5 \beta \text{PR} \quad \textbf{(4.12)}$$

The last term in Eq. (4.12) is a constant and does not depend on the design variables.

The values of q and c_B giving the minimum total (capital plus operating) cost can now be determined from Eq. (4.12). The volume and the feed rate of B are then determined from Eqs. (4.10). It is clear from inspection of Eq. (4.12) that a minimum cost exists. The cost becomes infinite as $q \to 0$ and as $q \to \infty$; hence, an intermediate flow rate will give the minimum cost. Similarly, the cost becomes infinite for both $c_B \to 0$ and $c_B \to \infty$, indicating an optimal intermediate value of c_B. The actual values will reflect trade-offs between the various contributions to the total cost, as reflected by the marginal cost factors $\{\lambda_i\}$. In this simple example, the optimum can be determined by setting $\partial \text{Cost}/\partial q$ and $\partial \text{Cost}/\partial c_B$ simultaneously to zero, yielding two additional design equations to be solved simultaneously with Eqs. (4.10):

$$\frac{\partial \text{Cost}}{\partial q} = \frac{\lambda_1 \text{PR}}{k(\text{FR} - \text{PR})} \frac{1}{c_B} + \lambda_2 - \frac{\lambda_3(\text{FR} - \text{PR})}{q^2} + \lambda_5 c_B = 0 \quad \textbf{(4.13a)}$$

$$\frac{\partial \text{Cost}}{\partial c_B} = -\frac{\lambda_1 \text{PR}}{k(\text{FR} - \text{PR})} \frac{q}{c_B^2} + \lambda_4 + \lambda_5 q = 0 \quad \textbf{(4.13b)}$$

4.4 TRANSIENTS AND TIME CONSTANTS

The well-stirred reactor provides a nice means of examining some of the factors that affect system dynamics. It is easier to consider the case of an irreversible reaction

$$A \rightarrow \text{products} \tag{4.14}$$

with a *first-order* reaction rate,

$$r = kc_A \tag{4.15}$$

Such a reaction scheme would rarely occur, except in nuclear fission, but it may approximate other systems in which all reactants but one are present in such excess that their concentrations change negligibly and they can be treated like inerts.

The flow rates are taken to be constant; with the usual linear assumption about the density, the volume is therefore constant, and Eq. (4.3a) for species A becomes

$$V \frac{dc_A}{dt} = q(c_{Af} - c_A) - kVc_A \tag{4.16}$$

We suppose that the temperature does not change, so that k is a constant. It is sufficient for illustrative purposes to take c_{Af} as constant as well. Finally, it is conventional to define the *mean residence time*, θ, as

$$\theta = V/q \tag{4.17}$$

θ can be shown by probabilistic arguments to be the mean time that a fluid particle spends in a perfectly mixed vessel, and it is determined entirely by hydraulics.

The transient solution to Eq. (4.16), given a concentration c_{A0} in the tank at $t = 0$, is

$$c_A(t) = c_{A0} \exp[-(1 + k\theta)t/\theta]$$
$$+ [c_{Af}/(1 + k\theta)]\{1 - \exp[-(1 + k\theta)t/\theta]\} \tag{4.18}$$

As $t \rightarrow \infty$, the concentration goes to a steady-state value, equal to $c_{Af}/(1 + k\theta)$. The response time is scaled with the *time constant*,

4.5 MULTIPLICITY AND INSTABILITY

$$\tau = \frac{\theta}{1 + k\theta} \tag{4.19}$$

The steady state is reached after a transient time equal to approximately three time constants. Real (clock) time is irrelevant; the system defines the appropriate measure for time.

In Eq. (4.15), k has dimensions of reciprocal time, so k^{-1} defines the time scale over which chemical reaction occurs. $k\theta$ is therefore a ratio of two time scales that characterize the system: the hydraulic time scale, θ, and the reaction time scale, k^{-1}. If $k\theta \ll 1$ (fast reaction relative to hydraulics), then $\tau \approx \theta$ and the dynamics are determined by the hydraulics, with c_A varying approximately as $\exp(-t/\theta)$. If $k\theta \gg 1$ (reaction slow relative to hydraulics), then $\tau \approx k^{-1}$ and the dynamics are controlled by the chemical reaction, with c_A varying approximately as $\exp(-kt)$.

The important point here is that the system must be properly scaled. It is the system (or its model) that defines the appropriate units, and establishes what is fast or slow, or long or short. The scales will themselves depend on the magnitudes of phenomena and the way in which they interact.

4.5 CATALYTIC OXIDATION: MULTIPLICITY AND INSTABILITY

The oxidation of carbon monoxide (CO) to carbon dioxide (CO_2) is one of the functions of the automobile catalytic muffler. The kinetic rate of oxidation of carbon monoxide over a platinum catalyst has an unusual shape, as shown in Fig. 4.3. At low concentrations (where the data are poor) the rate is a monotonically increasing function of concentration, while it *decreases* at high concentration. This type of rate equation can lead to rather interesting behavior. We introduce it here and consider some consequences because similar behavior is characteristic of many systems of greater complexity, but with less unusual constitutive relations.

Let us suppose that the oxidation is carried out in a well-stirred tank containing a highly dispersed catalyst; a fluidized bed (see Section 2.3) approximates this situation. (This is *not* the reactor configuration that would be used in practice!) Because the catalyst is well dispersed, we will treat the chemical reaction as though it takes place at all points in the reactor, and not simply at the solid surface; this is obviously an approximation, but if the mean distance between catalyst particles is small relative to the reactor dimensions, then it should be acceptable. Because the reaction takes place in the gas phase, the reactants fill the entire volume, so V is a constant. If

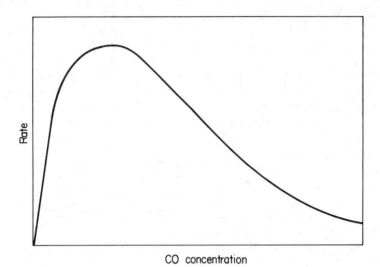

FIG. 4.3.
Qualitative representation of rate of reaction of CO over a platinum catalyst.

CO is present in the feed stream only in small quantities, then the volumetric flow rates in and out will be the same.*

With these physical assumptions, the equation for concentration of mass for CO is

$$\theta \frac{dc}{dt} = c_f - c - \theta r \qquad (4.20)$$

where $\theta = V/q$ and r is as given in Fig. 4.3. We have dropped the subscript on the concentration. At steady state, $dc/dt = 0$, and we can write

$$c_f - c = \theta r \qquad (4.21)$$

The left- and right-hand sides of Eq. (4.21) are plotted on Fig. 4.4. The equation is satisfied when the two curves intersect. *Note that as many as*

---

*This last restriction is necessary because of a consequence of the ideal gas equation, Eq. (3.7), or any other applicable gas constitutive equation. The reaction $CO + (1/2)O_2 \rightarrow CO_2$ reduces the total number of moles, so the effluent has a lower density than the feed at equal pressure and temperature. (The temperatures will not be the same either, in reality, but let us avoid this additional complication.) This density change will be negligible if the CO level in the feed is small, say below one percent.

4.5 MULTIPLICITY AND INSTABILITY

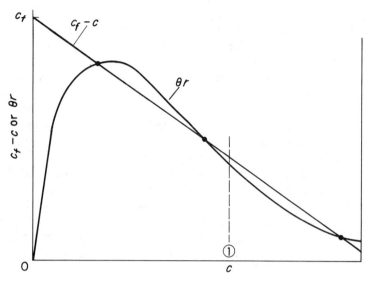

FIG. 4.4.
Graphical solution of Eq. (4.21).

three intersections are possible! For very large θ, only one intersection is possible, corresponding to a low CO concentration in the effluent (high conversion). For very small θ, there is also only one intersection, in this case at a CO concentration that is close to that of the feed (low conversion). At intermediate values, however, high, low, and intermediate conversion solutions are all possible.

The idea that nonlinear systems might admit multiple solutions is not new; everyone is familiar with the example $X^2 = 1$, with solutions $+1$ and -1. Students are usually taught to discard all but one solution on physical grounds—concentrations and absolute temperatures cannot be negative, for example. But here we have three solutions, all of which are physically reasonable. Two of the three are, in fact, physically *possible*, and the one that will be realized will depend on the manner in which we start the system up. Thus, it will be necessary to consider the dynamics in order to establish at which steady state the system will operate.

There is a more-or-less intuitive argument that can be used to establish in a reasonably convincing manner that the intermediate steady state is not one that the system can maintain. The left-hand side of Eq. (4.21) is the net rate at which reactant enters by flow (when multiplied by the constant flow rate), while the right-hand side is the rate of removal by chemical reaction. Suppose the system is somehow moved to a slightly higher concentration than the steady state, say point 1 in Fig. 4.4. The removal rate is below the net rate of inflow, so the concentration will tend to increase, driving the system to a still higher concentration. Similarly, if we move to a slightly lower concentration, the removal rate is greater than the net rate of inflow, so the concentration will tend to decrease even more. The conclusion is that any infinitesimal perturbation will tend to drive the system away from the intermediate steady state. Infinitesimal perturbations must always exist in any real physical system, so the intermediate steady state is *unstable* and cannot be maintained in practice without some sort of control system (which would, of course, change the process and the equations).

The same argument, if applied to the high and low conversion steady states, would seem to suggest that they are stable. The argument works in this case, but the same logic fails for processes for which the order of the differential equation is greater than first, so such reasoning must be used with the greatest caution. We shall return to this point subsequently.

It is perhaps instructive to look at the transient behavior analytically over a portion of the concentration range. The rate expression varies approximately as c^{-1} to the right of the maximum. If we then set

$$r = kc^{-1} \qquad (4.22)$$

recognizing that this form is not even qualitatively correct at low concentrations, the dynamical response is given by the equation

$$\theta \frac{dc}{dt} = c_f - c - \frac{k\theta}{c} \qquad (4.23)$$

The middle and upper steady states, denoted c_s, are then given by

$$c_s = \frac{c_f \pm \sqrt{c_f^2 - 4k\theta}}{2} \qquad (4.24)$$

The middle and upper steady state no longer exist (c_s becomes complex) for $\theta > c_f^2/4k$, which is therefore a uniqueness criterion (to the extent that Eq. (4.22) approximates the real rate expression beyond the maximum) for the high conversion steady state.

It is conventional (and useful) in analyzing dynamics to use the steady state as the frame of reference. Thus, we define a new dependent variable,

$$\xi = c - c_s \qquad (4.25)$$

and Eq. (4.23) becomes

$$\theta \frac{d\xi}{dt} = c_f - c_s - \xi - \frac{k\theta}{c_s + \xi}$$

$$= \left(c_f - c_s - \frac{k\theta}{c_s}\right) - \xi - k\theta\left(\frac{1}{c_s + \xi} - \frac{1}{c_s}\right)$$

$$= -\xi\left(1 - \frac{k\theta}{c_s^2 + c_s\xi}\right) \tag{4.26}$$

The first grouping in parentheses in the second line of Eq. (4.26) sums to zero because of the steady-state relation. For sufficiently small values of ξ, we can neglect $c_s\xi$ relative to c_s^2, so for small ξ we have, approximately,

$$\theta \frac{d\xi}{dt} = -\xi\left(1 - \frac{k\theta}{c_s^2}\right) \tag{4.27}$$

The solution to this equation is

$$\xi = \xi_0 \exp\left[-\left(1 - \frac{k\theta}{c_s^2}\right)t/\theta\right] \tag{4.28}$$

Thus, for $k\theta/c_s^2 < 1$, corresponding to the high concentration, low conversion steady state, $\xi(t)$ goes to zero at long times and c returns to the steady state (as long as ξ_0 is much smaller in magnitude than c_s, so that the approximation in passing from Eq. (4.27) to (4.28) is valid). For $k\theta/c_s^2 > 1$, which is the middle steady state, $\xi(t)$ grows without bound for arbitrarily small ξ_0. Thus, the system will always move away from this steady state, even from starting concentrations that are arbitrarily close, and it is absolutely unstable. The fact that $\xi(t)$ moves away from ξ_0 and from zero is the important point here, not that it grows without bound; clearly, at some point in this growth process the assumption that ξc_s is small compared to c_s^2 will fail, and Eq. (4.27) will no longer describe the physical process.

4.6 BIOLOGICAL AND OTHER AUTOCATALYTIC SYSTEMS

An autocatalytic reaction is one in which a reactant also appears as a product; the autocatalytic analogue to Eq. (4.1) is

$$A + \beta B \rightarrow \gamma B + \text{other products}, \quad \gamma > \beta \tag{4.29}$$

4 MODELING REACTORS: CONSERVATION OF MASS

The primary reaction in the activated sludge process, Section 2.6, is of this type, where A represents the biodegradable material and B the microorganism. Because of our subsequent interest in this latter process, it is useful to see what the consequences are of this small change in the reaction scheme.

If we are really considering a biological reactor, then the reaction mixture is a multiphase slurry of wastewater, suspended microorganisms, and air bubbles (to provide oxygen). As in the case of the fluidized bed reactor, we typically treat the mixture as though it were homogeneous, with reaction taking place at each point in the liquid slurry. The overall mass balance is unchanged, as is the mass balance for species A, Eq. (4.3a). The reaction stoichiometry now requires that for each mole of A that disappears, β moles of B disappear but γ moles be formed; hence, there is a net appearance of $(\gamma - \beta)$ moles of B, and we have a molar rate of *formation* equal to $(\gamma - \beta)r$. Conservation of mass for species B therefore becomes, in place of Eq. (4.3b),

$$\frac{d}{dt} c_B V = q_f c_{Bf} - q c_B + (\gamma - \beta) V r \qquad (4.30)$$

The processing problem of greatest interest to us is shown schematically in Fig. 4.5. There is no B in the external feed stream. A fraction f of the mass of species B in the reactor effluent is separated and recycled to the reactor, however, so

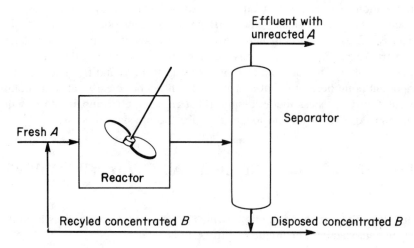

FIG. 4.5.
Schematic of a reactor for an autocatalytic reaction with product recycle.

4.6 BIOLOGICAL AND OTHER AUTOCATALYTIC SYSTEMS

$$q_f c_{Bf} = f q c_B \tag{4.31}$$

We assume for simplicity that r is given by the bilinear form, Eq. (4.7). At steady state ($d/dt = 0$), Eqs. (4.4a) and (4.30) then become

$$q_f c_{Af} - q c_A - V k c_A c_B = 0 \tag{4.32a}$$

$$q c_B (f - 1) + (\gamma - \beta) V k c_A c_B = 0 \tag{4.32b}$$

We first note that $c_B = 0$, $c_A = c_{Af}$ *is always a solution to Eqs.* (4.32). Since we are presumably interested in solutions for which $c_B \neq 0$, we must define the range of such solutions. We are always confronted with the fact that, *if a solution with* $c_B > 0$ *exists, the process equations admit two physically possible solutions* ($c_B > 0$, $c_B = 0$) *and process transient behavior may determine which state is actually attained.*

For $c_B \neq 0$, Eq. (4.32b) can be solved for c_A:

$$c_A = \frac{q(1-f)}{(\gamma - \beta) V k} \tag{4.33}$$

Note that f must be strictly less than unity; for $f = 1$, $c_A = 0$, in which case $c_B \to \infty$ in order to satisfy Eq. (4.33a). This is consistent with our physical understanding. If all of the B produced is returned to the reactor, then the amount of B will continue to grow without bound as long as there is sufficient A fed to the reactor for the reaction to occur. This is really a transient situation, however, and steady-state equations have no meaning.

With Eq. (4.33), we can solve Eq. (4.32a) for c_B:

$$c_B = \frac{(\gamma - \beta) c_{Af}}{1 - f} - \frac{q}{V k} \tag{4.34}$$

(For simplicity we have made the linear density assumption and set $q_f = q$.) This is a difference between two positive numbers. Since we require $c_B > 0$, Eq. (4.34) places a limit on the reactor variables:

$$c_B > 0 \Rightarrow \frac{q}{V} < \frac{(\gamma - \beta) k c_{Af}}{1 - f} \tag{4.35}$$

q/V is known in the environmental and biochemical engineering literature as the *dilution rate;* it is the reciprocal of the residence time, θ. If the dilution rate exceeds the bound set by Eq. (4.35), then the only physically meaningful solution that can exist for Eqs. (4.32) is $c_A = c_{Af}$, $c_B = 0$. Such a situation is known as *washout*. Equation (4.35) demonstrates the need for adequate flow control in autocatalytic systems, since excursions in flow rate or volume leading to long-term violations of Eq. (4.35) could take the system irreversibly to the washout state.

4.7 BATCH AND TUBULAR REACTORS

The batch reactor, which is often used for experimental determination of rates, is a well-stirred reactor for which $q_f = q = 0$. With the usual simplifying assumptions about the density, this implies that $dV/dt = 0$, $V =$ constant, and for a single reaction Eqs. (4.4) simplify to

$$\frac{dc_A}{dt} = -r, \frac{dc_B}{dt} = -\beta r, \ldots \quad (4.36)$$

Consider now a totally different reactor configuration, shown schematically in Fig. 4.6, which is also in common use. The reactants flow through a long tube at a constant mean linear velocity v. There is reasonable mixing across the radial plane, but little mixing in the axial direction; i.e., any axial mixing is on a length scale that is negligible relative to the reactor length. If we mark the fluid over a small spatial region Δz with a tracer (color it blue), that fluid element will retain its integrity as it passes through the reactor. Thus, it is in effect a little batch reactor that is moving at velocity v. Since the batch reactor equations do not depend on the size of the control volume, Δz here is arbitrary and can be as small as we wish.

Use of a control volume that moves with the fluid is known in the fluid mechanics literature as a Lagrangian description. We can convert to an Eulerian description, or a fixed laboratory coordinate system, by noting that the time required to travel distance dz is vdt, so we may replace dt with dz/v, and Eqs. (4.36) become

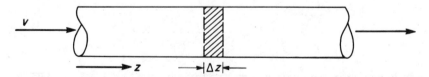

FIG. 4.6.
Differential control volume in a tubular system with no longitudinal mixing.

4.7 BATCH AND TUBULAR REACTORS

$$v\frac{dc_A}{dz} = -r, \; v\frac{dc_B}{dz} = -\beta r, \ldots \tag{4.37}$$

(This assumes, of course, that the system is at steady state when viewed from a fixed laboratory frame.) Thus, if we have a description of behavior in a batch reactor, then it can be carried over immediately to a tubular reactor as long as axial dispersion is unimportant and no significant radial gradients are expected. This concept will be of use to us in looking at some published modeling results in the next chapter.

5

MODELING REACTORS: CONSERVATION OF ENERGY

5.1 INTRODUCTION

The principle of conservation of energy is a seemingly elementary concept, but it has proved to be particularly elusive in applications. This is less surprising than it might seem when one examines development of the principle as described by historians of science. The application of energy conservation to the modeling of physical systems involving phase change or chemical transformation has been done incorrectly so often that there are examples of textbooks with sections devoted to demonstrating the importance of terms that are in error. Every textbook on process dynamics and control available at the time of writing derives the energy equation for a flowing, reacting system incorrectly.

5.2 THERMODYNAMIC VARIABLES

The principle of conservation of energy states that the rate of change of energy in the control volume equals the rate at which energy enters the control volume, minus the rate at which energy leaves. Energy enters or leaves by flow, by heating or cooling at the boundaries of the control volume, and

5.2 THERMODYNAMIC VARIABLES

by performance of work. One often sees terms purporting to account for "generation of energy (or heat)" through chemical reaction or phase change; no such terms exist.

Proper application of the principle of conservation of energy requires the use of some basic thermodynamic concepts. We shall touch on these, but a complete treatment is beyond our scope. The reader who is new to these ideas is cautioned that the application of thermodynamic concepts is a delicate enterprise. We must emphasize that there is no way that the proper use of thermodynamics can be avoided when dealing with the energy of a system. We have seen many attempts to bypass rigor and substitute "intuition." The end result is usually an incorrect equation, with all of the consequences that can follow. Unfortunately, thermodynamics of the type that we require to account for transitions of phase or species is not covered in the curricula of most disciplines that produce "modelers."

First, there are many forms of energy. *Kinetic* (KE) and *potential* (PE) *energy* are familiar concepts from mechanics; the former (on a unit mass basis) is $v^2/2$, while the latter is gh, where v is the magnitude of the velocity, h the elevation above an arbitrary datum, and g the gravitational acceleration. *Internal energy* can be thought of as the energy associated with the internal state of a body, as determined by the temperature, pressure, composition, and perhaps some other internal variables; at a molecular level, the internal energy reflects the state of molecular motion, including an intramolecular contribution from atomic association. The symbol U will be used for internal energy. We will usually not be concerned with additional forms of energy, and we will take the total energy to be the sum of these three energy forms:

$$\text{Total energy} = KE + PE + U \qquad (5.1)$$

Energy per unit mass will be denoted with an underbar:

$$\text{Energy/mass} = \underline{KE} + \underline{PE} + \underline{U} \qquad (5.2)$$

We will often be using a molar measure of concentration. Quantities per unit mole will be denoted with an undertilde:

$$\text{Energy/mole} = \underset{\sim}{KE} + \underset{\sim}{PE} + \underset{\sim}{U} \qquad (5.3)$$

In flowing systems, it is often more convenient to work in terms of the *enthalpy*, H, defined as

$$H = U + pV \qquad (5.4)$$

where V is the volume and p the mean pressure. Enthalpy per unit mass is then

$$\underline{H} = \underline{U} + p/\rho \tag{5.5}$$

and enthalpy per unit mole is

$$\utilde{H} = \utilde{U} + M_w p/\rho \tag{5.6}$$

where M_w is the average molecular weight. For an ideal gas, Eq. (3.7),

$$\utilde{H} = \utilde{U} + RT \tag{5.7}$$

Given the pressure-temperature-volume-composition constitutive equation for a material, the thermodynamic state is specified by the composition and two of the remaining three characterizing variables. Internal energy is usually defined in terms of the volume and temperature, while enthalpy is usually defined in terms of the pressure and temperature. The partial derivatives of internal energy and enthalpy with respect to temperature are known as *heat capacities;* these quantities can be measured in a calorimeter, and extensive data tabulations and correlations are available. The heat capacities at constant pressure are defined to be

$$c_p = \left.\frac{\partial \underline{H}}{\partial T}\right]_{p,\,\text{composition}} \tag{5.8a}$$

$$\utilde{c}_p = \left.\frac{\partial \utilde{H}}{\partial T}\right]_{p,\,\text{composition}} \tag{5.8b}$$

The heat capacities at constant volume are defined as

$$c_v = \left.\frac{\partial \underline{U}}{\partial T}\right)_{\rho,\,\text{composition}} \tag{5.9a}$$

$$\utilde{c}_v = \left.\frac{\partial \utilde{U}}{\partial T}\right)_{\rho,\,\text{composition}} \tag{5.9b}$$

It is worth noting in passing that the ideal gas can be approached from a thermodynamic framework, and defined as a material for which $\underline{U}$ is a function only of T. In that case, it follows from Eqs. (5.7) through (5.9) that

$$\utilde{c}_p = \utilde{c}_v + R \tag{5.10}$$

For liquids at moderate pressures and temperatures, c_p and c_v are nearly equal.

We will require one more thermodynamic quantity. Let n_i be the number of moles of species i contained in volume V. Define the *partial molar enthalpy*, $\widetilde{H}_i$, as

$$\widetilde{H}_i = \left.\frac{\partial H}{\partial n_i}\right)_{T, p, n_j = \text{constant}, \, j \neq i} \tag{5.11}$$

One of the consequences of the Gibbs-Duhem equation is that

$$H = \sum_i n_i \widetilde{H}_i \tag{5.12a}$$

or, equivalently,

$$\underline{H} = \frac{1}{\rho} \sum_i c_i \widetilde{H}_i \tag{5.12b}$$

where c_i is the molar concentration of species i. In an ideal solution, molecules of species i interact with molecules of all other species in the same way as with their own; thus $\widetilde{H}_i$ equals the enthalpy per mole of pure i, which we denote $\underline{H}_i$. In a nonideal solution, $\widetilde{H}_i \neq \underline{H}_i$, and there will be enthalpy changes associated with the mixing of different species.

5.3 CONSERVATION OF ENERGY

We are now ready to return to the reacting system considered in Chapter 4. The reaction

$$A + \beta B + \cdots \rightarrow \mu M + \nu N + \cdots \tag{4.1}$$

takes place in a well-stirred tank. (Recall that there is no loss of generality in taking the stoichiometric coefficient of A equal to unity.) Because of the well-mixed assumption, the entire tank is taken as the control volume. The species equations for conservation of mass are given by Eqs. (4.4):

$$\frac{dn_A}{dt} = \frac{d}{dt} c_A V = q_f c_{Af} - q c_A - Vr \tag{4.4a}$$

5 MODELING REACTORS: CONSERVATION OF ENERGY

$$\frac{dn_B}{dt} = \frac{d}{dt} c_B V = q_f c_{Bf} - q c_B - \beta V r \qquad (4.4b)$$

$$\frac{dn_M}{dt} = \frac{d}{dt} c_M V = q_f c_{Mf} - q c_M + \mu V r \qquad (4.4c)$$

$$\vdots$$

The principle of conservation of energy applied to this control volume is

$$\frac{d}{dt}(U + KE + PE) = \rho_f q_f (\underline{U}_f + \underline{KE}_f + \underline{PE}_f)$$

$$- \rho q (\underline{U}_e + \underline{KE}_e + \underline{PE}_e) + Q + W_T \qquad (5.13)$$

The first two terms on the right are the rates of convective flow of energy in and out, respectively. The subscript "e" denotes the effluent stream; despite perfect mixing, the energy of the effluent will be different from that in the tank. (This will not usually be true of the internal energy, since the temperature, density, and composition of the effluent will generally be the same as that in the tank.) Q is the rate of heat addition through the boundaries, typically from a heating or cooling coil or jacket; W_T is the rate at which work is done on the system (i.e., the power input).

Let us first look at the work term. Work is done on the system when fluid is forced in, and work is done by the system to expel the effluent stream; the rate of the former is $q_f p_f$, while the rate of the latter is $q p_e$, where p_f and p_e are the pressures just prior to the entrance and exit, respectively.* (Note that p_e is not the pressure in the effluent, but that at the exit. These may differ because of pressure drop across a valve.) It is convenient to separate out these work terms, and refer to the remaining work term as W_s, for *rate of shaft work*. We thus rewrite Eq. (5.13) as

$$\frac{dU}{dt} = \rho_f q_f \left(\underline{U}_f + \frac{p_f}{\rho_f} \right) - \rho q \left(\underline{U}_e + \frac{p_e}{\rho} \right) + Q + W_s \qquad (5.14)$$

We have dropped the kinetic and potential energy terms here, since they are usually unimportant if temperature changes of even a few degrees can occur.

---

*The total stress in the direction of flow should be used. This equals the pressure for Newtonian fluids, but may differ for materials like polymers. There will also be a term representing work against the environment because of volume change with time, but this will be negligible (except at high pressure) and can usually be neglected.

5.3 CONSERVATION OF ENERGY

This is, however, an approximation. From the definition of enthalpy, Eq. (5.5), we can write Eq. (5.14) as

$$\frac{dU}{dt} = \rho_f q_f \underline{H}_f - \rho q \underline{H}_e + Q + W_s \tag{5.15}$$

The pressure dependence of the enthalpy of a liquid is small, and enthalpy variations because of the change in pressure from the top to the bottom of the tank are negligible. The enthalpy of an ideal gas can be shown to be independent of pressure, and spatial pressure variations in the tank are negligible in any event for a gas. Thus, we can drop the subscript "e" on $\underline{H}_e$, since the enthalpy per unit mass is the same everywhere in the tank and will approximately equal the enthalpy of the effluent. The internal energy term on the left of Eq. (5.15) is usually rewritten in terms of enthalpy to give

$$\frac{dH}{dt} - \frac{d}{dt}(pV) = \rho_f q_f \underline{H}_f - \rho q \underline{H} + Q + W_s \tag{5.16}$$

The term $d(pV)/dt$ turns out to be surprisingly troublesome for many people. It is easily shown to be negligible in liquid systems, and is in fact identically zero if the volume and density are constant, so we neglect it relative to the remaining terms. It is often dropped in gas systems as well. This is obviously incorrect, and can lead to significant errors in transient modeling. We thus obtain our working equation *for liquid systems,*

$$\frac{dH}{dt} = \rho_f q_f \underline{H}_f(T_f) - \rho q \underline{H}(T) + Q + W_s \tag{5.17}$$

We have explicitly noted here the temperature at which each of the enthalpies on the right is to be evaluated, since this will be helpful later. Equation (5.17) is often quoted as the starting point in modeling, and referred to as the "enthalpy balance."*

---

*Enthalpy balance? Conservation of enthalpy? There is, of course, no such thing, and we have already seen that Eq. (5.17) is totally incorrect for gaseous systems. The "enthalpy balance" problem is actually worse, because another term is usually included to account for the "rate of enthalpy—or energy—generation because of chemical reaction"! This approach is a guaranteed path to disaster, providing correct answers only when they are already known, so that an appropriate series of approximations can be made to compensate for the incorrect starting point. Systems containing more than one phase cause particular problems for believers in "enthalpy balances."

5 MODELING REACTORS: CONSERVATION OF ENERGY

Equation (5.17) must now be expressed in terms of characterizing variables. The first step is to refer all enthalpies to the same temperature, which is most conveniently taken as the tank temperature. From the definition of c_p, Eq. (5.8a), we can write

$$\underline{H}_f(T_f) = \underline{H}_f(T) + \int_T^{T_f} c_{pf} \, dT \tag{5.18}$$

By $\underline{H}_f(T)$ we mean the enthalpy of the feed mixture, but evaluated at reactor temperature. c_p will be temperature dependent in general; for convenience we will take it to be approximately constant and write the integral as $c_{pf}(T_f - T)$; note that this approximation is not necessary. Equation (5.17) is now written

$$\frac{dH}{dt} = \rho_f q_f c_{pf}(T_f - T) + \rho_f q_f \underline{H}_f(T) - \rho q \underline{H}(T) + Q + W_s \tag{5.19}$$

H is a function of T, p, and the number of moles of all component species, $\{n_i\}$, and is thus an implicit function of time. We can thus write

$$\frac{dH}{dt} = \frac{\partial H}{\partial T}\frac{dT}{dt} + \frac{\partial H}{\partial p}\frac{dp}{dt} + \sum_i \frac{\partial H}{\partial n_i}\frac{dn_i}{dt} \tag{5.20}$$

The term $\partial H/\partial p$ can be shown to be negligible in most cases for liquid systems,* and it is identically zero for ideal gases. Thus, we will not consider it further here, but it can easily be retained if necessary. $\partial H/\partial T$ is simply $\rho V c_p$, from Eq. (5.8a), while $\partial H/\partial n_i$ is the definition of $\widetilde{H}_i$. Thus,

$$\frac{dH}{dt} = \rho V c_p \frac{dT}{dt} + \sum_i \widetilde{H}_i \frac{dn_i}{dt} \tag{5.21}$$

With Eqs. (4.4), we can write the sum in Eq. (5.21) as

$$\sum_i \widetilde{H}_i \frac{dn_i}{dt} = \rho_f \sum_i c_{if} \widetilde{H}_i - q \sum_i c_i \widetilde{H}_i$$

$$+ Vr[\mu \widetilde{H}_M - \widetilde{H}_A - \beta \widetilde{H}_B + \cdots] \tag{5.22}$$

---

*It is shown in thermodynamics textbooks that

$$\left.\frac{\partial H}{\partial p}\right)_{T, n_i} = V - T \left.\frac{\partial V}{\partial T}\right]_{p, n_i}$$

5.3 CONSERVATION OF ENERGY

Finally, using Eq. (5.12b), the term $q\Sigma c_i \widetilde{H}_i$ equals $\rho q \underline{H}$, while $\rho_f \underline{H}_f$ equals $\Sigma c_{if}\underline{H}_{if}$; combination of Eqs. (5.19) through (5.21) then becomes

$$\rho V c_p \frac{dT}{dt} = \rho_f q_f c_{pf}(T_f - T) - Vr[\mu\widetilde{H}_M - \widetilde{H}_A - \beta\widetilde{H}_B + \cdots]$$
$$+ Q + W_s + q_f \sum_i c_{if}(\widetilde{H}_{if} - \widetilde{H}_i) \quad (5.23)$$

The last term in Eq. (5.23) contains the difference between the partial molar enthalpy of each species at reactor temperature in two different mixtures, the feed and the reactor. If we have thermodynamically ideal solutions, then both are equal to the enthalpy per mole of the pure species i, and the term vanishes identically. The term is probably small in all instances relative to the enthalpy term multiplying Vr, and it is always neglected.* (Indeed, it is usually not even mentioned!)

The enthalpy term $\mu\widetilde{H}_M - \widetilde{H}_A - \beta\widetilde{H}_B + \cdots$ is the *enthalpy change of reaction*, often called the heat of reaction, and denoted $\Delta\underline{H}_R$; $\Delta\underline{H}_R$ is negative for an exothermic reaction and positive for an endothermic reaction. Values are tabulated, and enthalpies of reaction can be calculated from tabulated "heats of formation" and "heats of combustion." The enthalpy of reaction can be measured in a calorimeter experiment, and it is not necessary to have the individual partial molar enthalpies available. Our final form of the energy equation is therefore

$$\rho V c_p \frac{dT}{dt} = \rho_f q_f c_{pf}(T_f - T) + (-\Delta\underline{H}_R)Vr + Q + W_s \quad (5.24)$$

We emphasize again that this equation contains a large number of approximations, none of which should be serious for liquid systems. Neglect of the term $d(pV)/dt$ in Eq. (5.16) can cause large errors for a gas system. If the number of moles in the reactor is constant, and the gas can be taken to be ideal, then it follows from Eq. (5.10) that the correct equation for a gas phase system is obtained by replacing c_p in Eq. (5.24) (but *not* c_{pf}) with c_v, the heat capacity at constant volume. Recall also that c_{pf} has been taken as independent of temperature; if this is not true, then $c_{pf}(T_f - T)$ must be replaced with $\int c_{pf} dT$.

It suffices for our purposes to note that heat capacities and heats of reaction are available as functions of temperature and composition. We have

---

*This term is an enthalpy change on mixing. It is not clear how it could be measured, since any enthalpy change on mixing will be confounded by a simultaneous change because of the chemical reaction.

5 MODELING REACTORS: CONSERVATION OF ENERGY

already discussed the concentration dependence of the reaction rate constitutive equation in Chapter 4, with Eq. (4.6) as a typical form:

$$r = k_1 c_A{}^a c_B{}^b \cdots - k_2 c_M{}^m c_N{}^n \cdots \tag{4.6}$$

The coefficients will usually show a temperature dependence of Arrhenius, or Van't-Hoff, form:

$$k_i = k_{i0} \exp(-E_i/RT) \tag{5.25}$$

E_i has dimensions of energy/mole, and R is the gas constant. In expressions of this type, T is the *absolute* temperature (°K or °R); one sometimes sees $E_i/R(T + 273)$, in which case T is to be interpreted as °C. E_i/R will typically be of order 15,000 °K^{-1}.

The rate of heat transfer, Q, depends on the configuration used for heating or cooling. The simplest configuration to assume is that the reactor is jacketed and that the jacket fluid is a well-mixed liquid. Equation (5.24) is then applicable to the jacket as well, with appropriate identification of the variables. We will use a subscript "j" to denote "jacket," and assume that the jacket fluid is not reactive, so $r = 0$. We will also drop the shaft work term, since the jacket fluid will have a low viscosity and will not require substantial power input for mixing. The jacket equation is therefore

$$\rho_j V_j c_{pj} \frac{dT_j}{dt} = \rho_{jf} q_{jf} c_{pjf}(T_{jf} - T_j) + Q_j \tag{5.26}$$

If we assume that the only heat transfer is between the jacket and the reactor, with negligible loss from either to the surroundings, then clearly $Q_j = -Q$. Heat transfer rates usually vary linearly with heat transfer area and with the temperature difference, so we write

$$Q = -Q_j = hA(T_j - T) \tag{5.27}$$

A is the area available for heat transfer, and h is the *heat transfer coefficient*. Equation (5.27) is in reality a definition of h, since all of the other quantities can be measured. Many correlations exist for heat transfer coefficients.

Equation (5.26) can usually be simplified. The liquid volume in the jacket will not change, so $q_{jf} = q_j$; $c_{pjf} = c_{pj}$, since there is no reaction, and we have already taken c_{pjf} to be independent of temperature. There will be negligible error in taking $\rho_{jf} = \rho_j$. Thus, we can write

$$\rho_j V_j c_{pj} \frac{dT_j}{dt} = \rho_j q_j c_{pj}(T_{jf} - T_j) + hA(T - T_j) \tag{5.28}$$

5.4 BATCH AND TUBULAR REACTORS

The energy equation for a batch reactor is obtained directly from Eq. (5.24) by setting q_f to zero, as in Section 4.7. The derivation of the steady-state energy equation for tubular reactors without axial dispersion is essentially the same as the derivation of the mass balance in Section 4.7, but the heat transfer term must first be put in the appropriate form.

Let a_v represent the area available for heat transfer per unit volume of reactor. By comparison with Eq. (5.27) we have $A = a_v V$, and Eq. (5.24) can be written for a batch reactor as

$$\rho V c_p \frac{dT}{dt} = (-\Delta \underline{H}_R) V r + h a_v V (T_j - T) \tag{5.29}$$

(We have dropped the shaft work term for simplicity, since it will rarely be relevant in a tubular reactor.) Each term now contains V, which can be factored out. If we now repace dt with dz/v, then the equation for a tubular reactor becomes

$$\rho v c_p \frac{dT}{dz} = (-\Delta \underline{H}_R) r + \frac{4h}{d} (T_j - T) \tag{5.30}$$

We have made use of the fact that $a_v = 4/d$ for a tube of diameter d. The area per unit volume in the jacket, which is assumed to be a concentric outer tube of diameter d_j (Fig. 5.1), is $4d/(d_j^2 - d^2)$, where we have ignored the thickness of the wall. The corresponding equation for the jacket is then

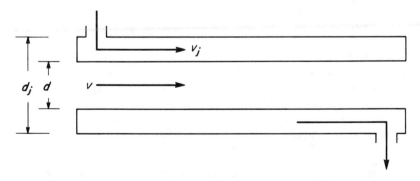

FIG. 5.1.
Schematic of a jacketed tubular reactor.

72   5 MODELING REACTORS: CONSERVATION OF ENERGY

$$\rho_j v_j c_{pj} \frac{dT_j}{dz} = \frac{4hd}{d_j^2 - d^2}(T - T_j) \tag{5.31}$$

Equations (5.30) and (5.31) are the equations for a shell-and-tube heat exchanger when $r = 0$. The flow is cocurrent if v and v_j have the same algebraic sign, and countercurrent if they have opposite algebraic signs.

We do need to emphasize the fact that this derivation of the tubular reactor equations is valid only for the steady state, and it assumes that radial mixing is so rapid that there are no radial concentration or temperature gradients; this follows from the assumption that the control volume $\pi d^2 dz/4$ is well mixed.

5.5   A CLASSIC (ALL TOO COMMON) BLUNDER

Equation (5.24), and its equivalent for more complex systems, is usually written incorrectly. (This is difficult to believe, but true. See pp. 4-23 and 4-24 in the fourth edition of Perry's *Chemical Engineers' Handbook* for a collection of incorrect design equations.) One sometimes sees

$$c_p \frac{d}{dt}\rho V T = \rho_f q_f c_{pf} T_f - \rho q c_p T + (-\Delta \underline{H}_R)\, Vr + Q + W_s$$

This form is readily shown to ignore a term $\rho_f q_f(c_p - c_{pf})T$, and is therefore equivalent to assuming that the heat capacities of feed and effluent are the same, which will often be a safe assumption. The equation is more commonly written in the incorrect form

$$\frac{d}{dt}\rho V c_p T = \rho_f q_f c_{pf}(T_f - T) + (-\Delta \underline{H}_R)rV + Q + W_s$$

or with some other right-hand side, but with the derivative term written $d(\rho V c_p T)/dt$. This form is grossly incorrect, and can lead to substantial errors; indeed, the effect of the incorrect term is large enough that one finds textbooks with examples demonstrating its importance!

Consider a case in which $\rho_f q_f = \rho q$, so ρV = constant, and in which c_p is temperature dependent. We can then write*

---

*There is a clear indication here that something is wrong without any consideration of the thermodynamics. The numerical value of the term $c_p + T\, dc_p/dT$ will be different, depending on whether the c_p-T functionality is expressed in °C or °K, using the same data. This is obviously physically unacceptable.

$$\frac{d}{dt}\rho V c_p T = \rho V \left(c_p + T \frac{dc_p}{dT}\right)\frac{dT}{dt}$$

Comparison with Eq. (5.24) shows that the coefficient of the temperature derivative contains an error that is proportional to the temperature. This term can be very important if the temperature is changing rapidly. In some calculations done by Khanna and Seinfeld for a methanation reactor, inclusion of the incorrect heat capacity term was found to lead to errors of more than 60% in the temperature rise across the bed, and 15% in overall conversion, for a heat capacity variation from 0.65 to 0.69. A long list of published applications using incorrect energy balances is contained in the Bibliographical Notes at the end of the chapter, including one from a text on modeling and simulation and another from a simulation of a coal gasifier by a commercial establishment. In these latter two cases the temperature derivatives are so large that the fundamental modeling error makes the computed results meaningless.

5.6 ORDER REDUCTION

Order reduction is one of the basic problems in dynamic modeling. The application of fundamental principles often leads to a set of model equations of high order, where the system order equals the sum of the orders of the individual differential equations. (An Nth order differential equation is equivalent to N first-order equations.) The number of variables in a high-order system can preclude ease of analysis. Further, a high-order system will usually have a wide range of time scales; such systems are called *stiff* in the numerical analysis literature, and are computationally difficult. The objective of order reduction is to obtain a low-order system with dynamical behavior that is essentially equivalent to the original high-order system, but that is more amenable to analysis and is computationally more tractable.

Order reduction is often done by estimation of the time scales associated with various parts of the process, and then making a pseudo-steady-state approximation for those process elements that respond quickly. We shall apply that procedure here in order to reduce the order of the chemical reactor equations, and in doing so we will see the limitations of this approach. A more systematic approach will be introduced in Chapter 9, but that approach is essentially limited to linear systems.

Our starting point is Eq. (5.28) for the jacket, which we can write in the form

$$\theta_j \frac{dT_j}{dt} = T_{jf} - T_j + \mathcal{H}(T - T_j) \qquad (5.32)$$

$$\theta_j = V_j/q_j, \qquad \mathcal{H} = hA/\rho_j q_j c_{pj} \qquad (5.33)$$

Here θ_j is the jacket residence time. The argument is sometimes made that the jacket residence time is much smaller than the reactor residence time, so jacket transients disappear rapidly relative to those in the reactor. Hence, jacket dynamics can be neglected, and we can set dT_j/dt to zero. The weakness in this argument is that it neglects the coupling through the heat transfer mechanism between jacket and reactor dynamics.

The essential structure comes out by rewriting Eq. (5.32) in the form

$$\frac{\theta_j}{1 + \mathcal{H}} \frac{dT_j}{dt} + T_j = \frac{\mathcal{H}}{1 + \mathcal{H}} T + \frac{1}{1 + \mathcal{H}} T_{jf} \qquad (5.34)$$

We thus see that the jacket dynamics are first order, with a time constant $\theta_j/(1 + \mathcal{H})$, with T and T_{jf} playing the role of time-dependent forcing functions. The jacket temperature will essentially reach steady state following a step change within about three time constants. Thus, if changes in the jacket feed and reactor temperatures take place over a time scale that is long relative to the jacket time constant, the jacket will respond dynamically on that time scale as though it were at steady state, and the time derivative can be neglected:

$$T_j \approx \frac{\mathcal{H}}{1 + \mathcal{H}} T + \frac{1}{1 + \mathcal{H}} T_{jf} \qquad (5.35)$$

(An alternate way of stating the same information is that the jacket is a low-pass filter.) We emphasize that Eq. (5.35) is valid only for transients that are slow on a time scale $\theta_j/(1 + \mathcal{H})$.

5.7 MULTIPLICITY

The pseudo-steady state approximation for the jacket enables reduction of the order of the reactor equations by one. If we consider the particular chemical reaction given by Eq. (4.14),

$$A \rightarrow \text{products} \qquad (4.14)$$

then the only species mass balance that we need consider is that for component A. If we further assume that $V = $ constant, $q = q_f$, then the system is second order, consisting of two first-order differential equations, one each

5.7 MULTIPLICITY

for the temperature and concentration of A. One advantage of reduction to a second-order system is that the system behavior can be inspected graphically.

The process can be analyzed from this point without any specific constitutive equation for the reaction rate, as long as the rate is a monotonically increasing function of c_A and of T, as will usually be the case. It is easier to obtain explicit expressions, however, if we take the rate to be first order, Eq. (4.15) with k given by Eq. (5.25):

$$r = k_0 c_A \exp\left(-\frac{E}{RT}\right) \tag{5.36}$$

The dynamical equations are then

$$V \frac{dc_A}{dt} = q(c_{Af} - c_A) - k_0 V c_A \exp\left(-\frac{E}{RT}\right) \tag{5.37}$$

$$\rho V c_p \frac{dT}{dt} = \rho q c_p (T_f - T)$$
$$+ (-\Delta \underset{\sim}{H_R}) k_0 V c_A \exp\left(-\frac{E}{RT}\right) + \frac{hA}{1 + \mathcal{H}} (T_{jf} - T) \tag{5.38}$$

Here, we have used Eq. (5.35) to eliminate T_j from the energy equation, and we have dropped the shaft work term for convenience, since it will rarely be important. We have also assumed for convenience that $\rho = \rho_f$, $c_p = c_{pf}$. We only consider the case in which q, q_j, T_f, and T_{jf} are constant in time.

We shall discuss scaling in detail in a later chapter, but it is helpful to introduce the basic notion here. It is always best to scale dependent variables such that the new variables are of order unity. Such a scaling simplifies approximation, and it also helps in visualization. The concentration and temperature are logically scaled with respect to the feed values, for example:

$$x = c_A/c_{Af}, \qquad y = T/T_f \tag{5.39a,b}$$

Time is most conveniently scaled with respect to the mean hydraulic residence time,

$$\tau = t/\theta = qt/V \tag{5.39c}$$

We have already seen in Section 4.4 that this scaling of time corresponds to use of the process time constant only in the case of no reaction, but θ is the most convenient single measure of time that we have for the second-order system. We thus obtain the dimensionless equations

$$\frac{dx}{d\tau} = 1 - x - \alpha x e^{-\gamma/y} \tag{5.40}$$

$$\frac{1}{1+\delta}\frac{dy}{d\tau} = \phi - y + \alpha\beta x e^{-\gamma/y} \tag{5.41}$$

The dimensionless parameters α, β, γ, δ, and ϕ are defined as follows:

$$\alpha = \frac{k_0 V}{q} = k_0\theta, \qquad \beta = \frac{(-\Delta H_R)c_{Af}}{\rho c_p T_f (1+\delta)}$$

$$\gamma = \frac{E}{RT_f}, \qquad \delta = \frac{hA}{\rho q c_p (1+\mathcal{H})}, \qquad \phi = \frac{1 + \delta T_{jf}/T_j}{1+\delta} \tag{5.42}$$

β is positive for an exothermic reaction. δ goes to zero for an adiabatic reactor, in which there is no heat transfer area, and ϕ then goes to unity.

Many people are uncomfortable with dimensionless equations, which they see as a formulation that moves the mathematical description away from the physical problem. We believe that the advantage of working with a properly scaled, dimension-free set of equations far outweighs any disadvantages resulting from the need to interpret the parameters in Eq. (5.42).

The right-hand sides of Eqs. (5.40) and (5.41), considered as a single vector differential equation, satisfy a Lipschitz condition, which ensures that the transient solution will be unique for arbitrary initial values x_0, y_0. In order to determine the transient conditions that we might expect, both for this process and for other similar nonlinear systems, it is best to start by examining the steady-state properties of the equations.

The steady state is obtained by setting the time derivatives to zero in Eqs. (5.40) and (5.41):

$$0 = 1 - x - \alpha x e^{-\gamma/y} \tag{5.43}$$

$$0 = \phi - y + \alpha\beta x e^{-\gamma/y} \tag{5.44}$$

We obtain a linear relationship between x and y by multiplying Eq. (5.43) by β and adding to Eq. (5.44):

$$\beta x = \beta + \phi - y \tag{5.45}$$

5.7 MULTIPLICITY

Since the steady-state concentration is bounded from above by the feed concentration and from below by zero, we must have

$$0 \leq x \leq 1 \tag{5.46}$$

In that case, Eq. (5.45) implies that

$$\phi \leq y \leq \phi + \beta \tag{5.47}$$

We can combine Eqs. (5.44) and (5.45) to obtain a single equation for dimensionless temperature, which is usually written as

$$\frac{1}{\alpha}(y - \phi) = F(y) \equiv (\beta + \phi - y)\, e^{-\gamma/y} \tag{5.48}$$

The function $F(y)$ has the shape of the solid line in Fig. 5.2, with an inflection and a maximum at intermediate values of y. The dashed lines in Fig. 5.2 are the function $(y - \phi)/\alpha$ for various values of α. According to Eq. (5.48), the steady state will occur at those values of y for which the two lines intersect. It is clear by inspection that there is a range of values of α for which three intersections are possible, corresponding to three steady-state solutions to the reactor equations having physically permissible concentration and temperature.

This possible multiplicity of solutions is similar to what we observed in Sections 4.4 and 4.5, where only conservation of mass was included. In those examples the nonlinear behavior that gave rise to multiplicity came from rather specialized kinetics. Multiplicity is a consequence of the simplest rate constitutive equation when the coupling between mass and energy conservation is included, and hence should always be anticipated. Since α is proportional to the residence time, θ, we see from Fig. 5.2 that we can expect a single steady state for a sufficiently large or small residence time; the former will correspond to a high temperature (y close to $\phi + \beta$) and a large conversion (small x), while the latter will correspond to a low temperature (y close to ϕ) and small conversion (x close to 1).

The behavior predicted by the model is illustrated in Fig. 5.3, with data of Vejtasa and Schmitz on the liquid phase reaction

$$2Na_2S_2O_3 + 4H_2O_2 \rightarrow Na_2S_3O_6 + Na_2SO_4 + 4H_2O$$

This reaction is a bit more complicated than Eq. (4.14), but the qualitative behavior will be unchanged. The data were taken in an adiabatic ($\delta = 0$) continuous flow stirred reactor. Vejtasa and Schmitz varied the residence

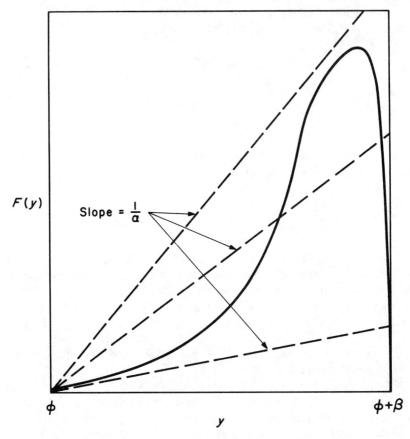

FIG. 5.2.
Graphical solution of Eq. (5.48).

time for fixed feed conditions and measured the steady-state temperature. At small and large residence times there was a unique steady-state operating temperature. For θ between 7 and 18 sec, however, two distinct steady operating temperatures were obtained, depending on startup conditions.

The data show two steady states at intermediate residence times, but the model predicts three. The middle steady state is readily shown to be unstable, in that the process moves away from the steady state with starting conditions that are arbitrarily close to the steady-state value. Thus, since arbitrarily small disturbances will always enter the process, the middle steady state cannot be maintained in the absence of a control system.

The physical argument that is usually employed to demonstrate instability of the middle steady state dates to a 1918 qualitative treatment of the ammonia synthesis reactor by Liljenroth. The function $F(y)$ on the right

5.7 MULTIPLICITY

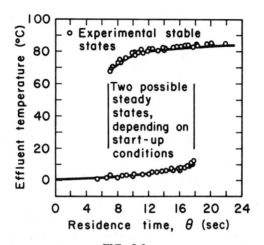

FIG. 5.3.
Experimental reactor temperature as a function of residence time (Vejtasa and Schmitz, 1970, copyright American Institute of Chemical Engineers, reproduced with permission).

of Eq. (5.48) represents the rate of enthalpy change by reaction (the term $-\Delta \underline{H}_R r$ in the dimensional equation), while the linear term on the left represents the net rate of removal by heat transfer and flow. If the rate of enthalpy change exceeds the rate of removal, then the temperature will increase for an exothermic reaction, causing the reaction rate to increase, and hence to increase the rate of enthalpy change by reaction even more, causing a further temperature increase, etc. A similar argument applies, but with a temperature decrease, if the rate of enthalpy change by reaction is less than the removal rate. We see from Fig. 5.2 that this is the situation in the neighborhood of the middle steady state, where the slope of the function $F(y)$ exceeds that of the removal rate. The situation is opposite at the high and low temperature steady states. This analysis appears to be (and is, in fact) equivalent to having a negative spring constant in a mass-spring-dashpot system, so that any deviation from equilibrium drives the system even further away.

The equivalent quantitative argument is instructive. We suppose that the system is perturbed slightly from the steady state in a very special way, so that Eq. (5.45) is still satisfied. In that case, Eq. (5.41) for the temperature will become

$$\frac{1}{1+\delta}\frac{dy}{d\tau} = \phi - y + \alpha F(y) \tag{5.49}$$

It is convenient to refer temperatures to the steady state, which we denote by y_s:

$$y = y_s + \eta \qquad (5.50)$$

In that case, Eq. (5.49) can be written

$$\frac{1}{1+\delta}\frac{d\eta}{d\tau} = \phi - y_s - \eta + \alpha F(y_s + \eta)$$

$$= -\eta + \alpha\,[F(y_s + \eta) - F(y_s)] \qquad (5.51)$$

Here we have made use of Eq. (5.44) and the fact that $dy_s/d\tau = 0$. For sufficiently small η, we may write $F(y_s + \eta) - F(y_s) \approx F'(y_s)\eta$, where $F'(y_s)$ denotes the derivative evaluated at y_s, with an error that goes to zero much faster than η. In that case,

$$\frac{1}{1+\delta}\frac{d\eta}{d\tau} = -\eta\,[1 - \alpha F'(y_s)] \qquad (5.52)$$

The solution to Eq. (5.25) is

$$\eta = \eta_0 \exp\{-[1 - \alpha F'(y_s)](1 + \delta)\tau\} \qquad (5.53)$$

Thus, for $\alpha F'(y_s) < 1$, corresponding to the middle steady state, solutions to the equation will grow without bound for arbitrarily small η_0, and the steady state is unstable.

The converse argument cannot be applied at the high and low temperature steady states. Equation (5.45) represents a very special transient. The fact that the process is unstable at the middle steady state when experiencing that transient is sufficient to prove instability, since that transient could occur, and we need only demonstrate one unstable event. This tells us nothing about the other steady states, though, because even though this particular perturbation does not correspond to instability, we cannot be sure that some other perturbation will not cause the process to run away. (The equivalent observation in a mass-spring-dashpot system is that, while a negative spring constant guarantees instability, a positive spring constant does not guarantee stability.) The physical argument leading to the slope condition fails in an equivalent manner, since we were applying a steady-state argument to a dynamical process and were therefore assuming a special disturbance.

5.8 TRANSIENT BEHAVIOR

The simplest model of a continuous flow well-mixed chemical reactor that includes energy conservation predicts three steady states over a finite range of operating variables. We have shown that one of these steady states cannot be realized because it is unstable following an arbitrarily small disturbance; the two remaining steady-state operating points are observed experimentally, and we shall show subsequently that the model predicts that they are stable to small disturbances. It is clear that there are some interesting and complex phenomena that are to be expected during transients. If either of two steady states can be reached for identical operating conditions, then the ultimate steady state will depend on the conditions during startup. Furthermore, very small changes in operating conditions could lead to extremely large changes in the state of the system.

This last point is nicely illustrated with Fig. 5.3. Suppose a reactor is designed to operate at $T = 67°$ C, with $\theta = 7$ sec. A slight increase in flow rate would reduce the residence time only slightly, but the temperature following the transient would fall to $3°$ C. If q were then slightly decreased, the steady-state temperature would move along the lower branch of the temperature-residence time curve. Conversely, a system designed to operate at a temperature of about $12°$ C at $\theta = 18$ sec would "run away" to a temperature in excess of $80°$ C with a slight increase in θ. Runaway is a potential problem in industrial reactors. It has been shown, for example, that the optimal operating point for a TVA-type ammonia reactor is close to unstable operation. The reactor for manufacture of phthalic anhydride may have five steady states, three of which are stable, with the optimal operating point at the middle steady state, and with only a small region separating the optimal point from the neighboring unstable states.

It is worth noting at this point some of the other types of transient behavior that can be observed. Figure 5.4 shows the results of an experiment by Baccaro, Gaitonde, and Douglas on the liquid phase reaction

$$CH_3COCl + H_2O \rightarrow CH_3COOH + HCl$$

in a well-stirred continuous flow reactor with cooling. The experiment was run under conditions corresponding to a unique steady state, but steady state could never in fact be obtained. Rather, the effluent cycled continuously with a fixed period, despite constant feed conditions. Such self-induced cycling, which is known as a *limit cycle,* is predicted by Eqs. (5.40) and (5.41) for some parameter values. Reactor cycling has often been observed in industrial scale systems. Cyclic reactions are known to occur in some biolog-

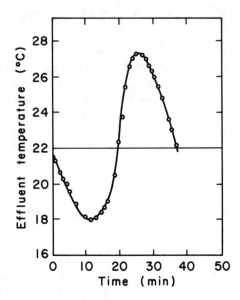

FIG. 5.4.
Temperature oscillations in a well-stirred reactor with steady feed (Baccaro, Gaitonde, and Douglas, 1970, copyright American Institute of Chemical Engineers, reproduced with permission).

ical systems, and may be the mechanisms for the "biological clocks" that govern the response of many organisms.

Some continuous reacting systems fail to achieve steady state, but the transient is nonperiodic and appears to be chaotic. Figure 5.5 shows data of Subramaniam and Varma from the reaction of CO and NO with oxygen carried out over a platinum-alumina catalyst in the presence of water vapor. The phenomena leading to this behavior are evidently more complex than those described by the simplest stirred tank equations, and Eqs. (5.40) and (5.41) do not predict chaotic response for any parameter values. Such behavior is nevertheless a property of relatively elementary mathematical structures of order at least three. Perhaps the most famous is the "Lorenz Attractor,"

$$\frac{dx}{dt} = -a(x - y) \qquad (5.54a)$$

$$\frac{dy}{dt} = -xz + cx - y \qquad (5.54b)$$

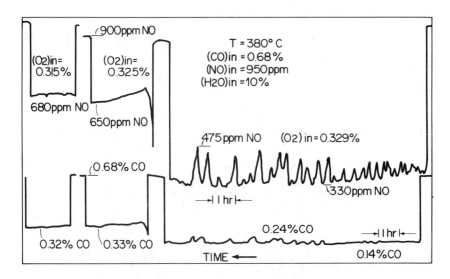

FIG. 5.5.
Concentrations of CO and NO when reacted with oxygen over a platinum-alumina catalyst in the presence of water vapor. Note the apparently chaotic response. (Subramaniam and Varma, 1982, copyright Gordon and Breach Science Publications, reproduced with permission).

$$\frac{dz}{dt} = xy - bz \tag{5.54c}$$

which shows a dynamical response resembling turbulence for certain parameter values. Kahlert and coworkers have demonstrated similar behavior in a simulation of a well-stirred reactor with two consecutive chemical reactions, the first exothermic and the second endothermic. Figure 5.6 shows some unpublished calculations by Caneba for a continuous flow stirred tank reactor with the single first-order reaction $A \rightarrow$ products in a parameter range where sinusoidal forcing of feed flow rate and coolant temperature causes an apparently turbulent response; the forcing substitutes for the third equation, and shows that the complex dynamics is a characteristic of the simplest reactor model. The study of such *strange attractors* is a relatively new field, and few general results are available, but it is clear that "turbulent" behavior is often to be expected. The ability to predict complex transients should serve as an effective model discriminator.

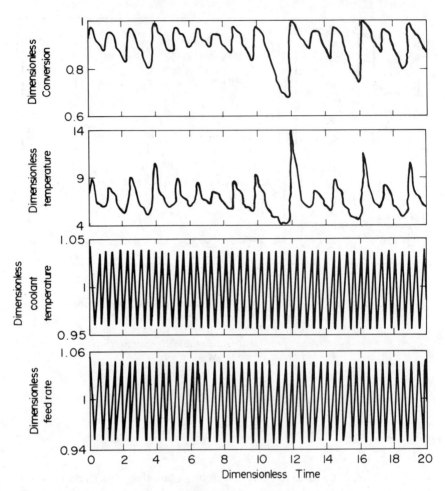

FIG. 5.6.
Conversion and temperature in a stirred-tank reactor with a single first-order reaction forced by sinusoidal variations in feed rate and coolant temperature (G. Caneba, unpublished).

5.9 FLUID CATALYTIC CRACKER (FCC) MODEL

We close this chapter with an example of the application of the modeling concepts developed thus far to the industrially important case of the fluid catalytic cracker, which was described in Section 2.3. A number of proprietary models of this process are known to exist. The best model available in the open literature seems to be one developed by Kurihara in a 1964

5.9 FLUID CATALYTIC CRACKER (FCC) MODEL

doctoral thesis, and it is that model that we describe here. The Kurihara model, with some modifications that have been made since by various authors, is generally recognized as describing the important dynamical features of an FCC, although it is probably not applicable to processes operating with the newest catalyst formulations.

5.9.1 The Problem

The reactive catalyst in the fluidized bed reactors is well mixed. Because of the small size of the catalyst particles, heat transfer is rapid, and solid and gas are assumed to be at the same temperature throughout the bed. Although the chemical reactions take place on the surface of the catalyst, the catalyst size is so small relative to the reactor dimensions that the reactions can be pictured as taking place in a homogeneous "fluid" whose properties are some average of those of the gas and the fluidized solid. The solid holdup in each reactor is assumed to be constant in the Kurihara model, although other authors have relaxed this assumption in considering control policies for the process. Thus, the reactors can be considered to be well-stirred continuous flow reactors of constant volume. Some of the cracking reaction takes place in the oil riser, and its volume is included as part of the cracking reactor.

It is important to keep Kurihara's goal in mind. He was concerned with the sluggish response of FCCs following process disturbances, and he wished to design an improved process control system using optimal control theory. Two process safety constraints needed to be satisfied in the regenerator: the regenerator temperature cannot exceed a maximum value, and the oxygen in the flue gas leaving the regenerator must be below some maximum. (The latter constraint is to protect the cyclone separator removing fine particles from the flue gas from excessive afterburning of carbon monoxide, which could cause damage to the materials of construction.) Kurihara was not concerned with the possible effect of product variations on downstream processing units or with the need of the operator to adjust the FCC operating characteristics to meet varying process requirements.

The major effort in developing the reactor models is in establishing the relevant chemistry and the reaction rate constitutive expressions. Kurihara used results from the published literature for his basic data, and we will take the rate expressions as given. We will also suppose, following Kurihara, that we are satisfied to consider the rate of conversion of gas oil and not the detailed product distribution. In that case, we need only focus on the carbon formation reactions in the cracking reactor and on the carbon removal reactions in the regenerator. As noted in Section 2.3, the formation reactions are endothermic, while the latter (combustion) are exothermic, leading to an overall thermal balance between the two reactors. We will use Kurihara's notation, which differs from the notation that we have employed thus far, in order to facilitate reading of the published literature. The notation is summarized in Table 5.1, together with typical values for each quantity.

5.9.2 Model Equations

Two different carbon species must be accounted for. The carbon that is deposited on the catalyst during the course of the cracking reactions is denoted *catalytic carbon*, with a symbol C_{cat}; C_{cat} is in weight percent, so the total mass of catalytic carbon in the cracking reactor is $H_{ra} C_{cat}/100$, where H_{ra} is the total mass of solids in the reactor. The remaining (*additive*) carbon is generally present in gas oil and is deposited without reaction. Carbon that has passed into or through the regenerator on spent or regenerated catalyst has a different chemical activity from the catalytic carbon; this species is identified as C_{rc} or C_{sc}, depending on whether the catalyst has been regenerated (i.e., is newly supplied to the reactor) or is spent. The mass balance on catalytic carbon in the cracking reactor is

$$H_{ra}\frac{dC_{cat}}{dt} = -60 R_{rc} C_{cat} + 50 R_{cf} \tag{5.55}$$

The factors 50 and 60 are a consequence of the system of units. R_{rc} is the mass flow rate of catalyst, so $R_{rc}C_{cat}$ is simply the rate at which catalytic carbon leaves the reactor. The activity is lost in the regenerator, so there is no inflow term. R_{cf} is the rate of formation of catalytic carbon. This term equals a rate per unit mass of solid multiplied by H_{ra}. The experimentally determined specific rate is proportional to total reactor pressure, which is a measure of the amount of gaseous carbon-containing reactants present, and is inversely dependent on the concentrations of both carbon species, with an Arrhenius temperature dependence:

$$R_{cf} = H_{ra} P_{ra} \frac{k_{cc}}{C_{cat}C_{rc}^{0.06}} \exp\left[-\frac{E_{cc}}{R(T_{ra} + 460)}\right] \tag{5.56}$$

k_{cc} and E_{cc} are constants.

The remaining reactor mass balance can be written for either the residual carbon or total carbon, and Kurihara chose the latter. Only catalytic carbon is formed, so the rate term is the same, and we have

$$H_{ra}\frac{dC_{sc}}{dt} = 60 R_{rc}(C_{rc} - C_{sc}) + 50 R_{cf} \tag{5.57}$$

TABLE 5.1.
Nomenclature and units for fluid catalytic cracker model. Typical values used by Kurihara for the simulations shown here (modified slightly by Nakano) are in parentheses.

Symbol	Quantity	Units
C_{cat}	Catalytic carbon on spent catalyst (0.875)	wt%
C_{rc}	Carbon on regenerated catalyst (0.6)	wt%
C_{sc}	Total carbon on spent catalyst (1.475)	wt%
D_{tf}	Density of total feed (7.0)	lb/gal
E_{cc}	Activation energy, carbon formation (18,000)	BTU/lb mole
E_{cr}	Activation energy, cracking (27,000)	BTU/lb mole
E_{or}	Activation energy, carbon burning (63,000)	BTU/lb mole
H_{ra}	Reactor catalyst holdup (60)	ton
H_{rg}	Regenerator catalyst holdup (200)	ton
k_{cc}	Reaction rate constant, carbon formation (8.59)	Mlb/hr psia ton (wt %)$^{-1.06}$
k_{cr}	Reaction rate constant, cracking (1.16 × 10^4)	Mbbl/day psia ton (wt%)$^{-1.15}$
k_{or}	Reaction rate constant, carbon burning (3.5 × 10^{10})	Mlb/hr psia ton
O_{fg}	Oxygen in flue gas (0.2)	mol%
P_{ra}	Reactor pressure (40)	psia
P_{rg}	Regenerator pressure (25)	psia
R_{ai}	Air rate (400)	Mlb/hr
R_{cb}	Coke burning rate	Mlb/hr
R_{cf}	Total carbon forming rate	Mlb/hr
R_{oc}	Gas-oil cracking rate	Mlb/hr
R_{rc}	Catalyst circulation rate (40)	ton/min
R_{tf}	Total feed rate (100)	Mbbl/day
S_a	Specific heat of air (0.3)	Btu/lb °F
S_c	Specific heat of catalyst (0.3)	Btu/lb °F
S_f	Specific heat of feed (0.7)	Btu/lb °F
T_{ai}	Air inlet temperature (175)	°F
T_{fp}	Feed preheater temperature (744)	°F
T_{ra}	Reactor temperature (930)	°F
T_{rg}	Regenerator temperature (1155)	°F
t	Time	hr
ΔH_{cr}	Heat of cracking (160)	Btu/lb
ΔH_{fv}	Heat of feed vaporization (60)	Btu/lb
ΔH_{rg}	Heat of regeneration (10700)	Btu/lb

5 MODELING REACTORS: CONSERVATION OF ENERGY

The derivation of the reactor energy balance assumes that the product of mass times heat capacity ($\partial H/\partial T$ in Eq. 5.20) is approximated by the mass of catalyst times catalyst heat capacity. The energy equation is then

$$S_c H_{ra} \frac{dT_{ra}}{dt} = 60\, S_c R_{rc}\,(T_{rg} - T_{ra})$$
$$+ \frac{7}{8} S_f D_{tf} R_{tf}(T_{fp} - T_{ra}) + \frac{7}{8}(-\Delta H_{fv})\, D_{tf} R_{tf} + \frac{1}{2}(-\Delta H_{cr}) R_{oc} \quad (5.58)$$

Equation (5.58) differs from Eq. (5.24) in a few important respects. First, the reactor is taken as adiabatic ($Q = 0$). There are two temperature difference terms because there are two feeds, one containing regenerated catalyst at T_{rg} and one containing gas oil at temperature T_{fp}. The gas oil enters as a liquid and is vaporized, so Eq. (5.18) must contain an additional term accounting for the difference between the enthalpy of the vapor and the enthalpy of the liquid feed; this is the source of the term containing the enthalpy of vaporization, ΔH_{fv}. R_{oc} is the overall rate of the cracking reaction, which is given by*

$$R_{oc} = \frac{1.75\, H_{ra} P_{ra} D_{tf} K_{cr}}{1 + H_{ra} P_{ra} K_{cr}/R_{tf}} \quad (5.59a)$$

$$K_{cr} = \frac{k_{cr}}{C_{cat} C_{rc}^{0.15}} \exp\left[-\frac{E_{cr}}{R(T_{ra} + 460)}\right] \quad (5.59b)$$

k_{cr} and E_{cr} are constants.

The regenerator mass and energy balance equations are, respectively,

---

*Appearance of a term proportional to the reactor residence time (H_{ra}/R_{tf}) in the rate expression is very bothersome, since the rate should depend only on local conditions (Compare Section 4.2). Comparison with Eq. (4.37) identifies the source of this term, and establishes that some order reduction has already been carried out. The gas oil reaction is believed to be second order in gas oil concentration. The reaction rate will therefore be proportional to $P_{ra} K_{cr} C_{Go}^2$, where C_{Go} is the gas oil concentration in any system of units. Kurihara visualized the gas as being in *plug flow* through a reactor at constant temperature and with a uniform (well-mixed) carbon distribution. The plug flow equation at steady state then requires that dC_{Go}/dz be proportional to $-C_{Go}^2$. Integration over the length of the reactor gives the change in gas oil concentration as $(H_{ra} P_{ra} K_{cr}/R_{tf})\,[1 + (H_{ra} P_{ra} K_{cr}/R_{tf})]$; multiplication by the feed rate, R_{tf}, gives the overall rate of conversion in precisely the form of Eq. (5.59a). Thus, the assumption has been made explicitly that the gas oil mass balance equation is at steady state. The justification is analogous to that used in Section 5.6. The residence time for gas oil in the reactor is of order seconds, while changes in carbon levels will occur over minutes or even hours. Thus, the oil gas conversion can be considered to be at steady state on the time scale of interest.

5.9 FLUID CATALYTIC CRACKER (FCC) MODEL

$$H_{rg} \frac{dC_{rc}}{dt} = 60 R_{rc} (C_{sc} - C_{rc}) - 50 R_{cb} \quad (5.60)$$

$$S_c H_{rg} \frac{dT_{rg}}{dt} = 60 S_c R_{rc} (T_{ra} - T_{rg}) + \frac{1}{2} S_a R_{ai} (T_{ai} - T_{rg})$$
$$+ \frac{1}{2} (-\Delta H_{rg}) R_{cb} \quad (5.61)$$

The two temperature difference terms in Eq. (5.61) reflect the separate catalyst and air feeds. ΔH_{rg} is the enthalpy change of the regeneration reactions, which is negative for the exothermic combustion process. (The negative sign is omitted in the published literature, and ΔH_{rg} is taken as a positive quantity. This is consistent with the practice for "heats of combustion," but contrary to the convention for "heats of reaction.") The carbon burning rate R_{cb} is given by*

$$R_{cb} = \frac{R_{ai} (21 - O_{fg})}{100 C_1} \quad (5.62a)$$

$$O_{fg} = 21 \exp \left\{ -\frac{H_{rg} P_{rg}/R_{ai}}{(10^6/4.76 R_{ai}^2) + 100/k_{or} \exp [-E_{or}/R(T_{rg} + 460)] C_{rc}} \right\}$$
$$(5.62b)$$

C_1 is a stoichiometric coefficient that reflects the selectivity of the combustion reaction between CO_2 and CO as products; a value of 2.0 used for calculations reflects a molar ratio of unity. O_{fg} is volume percent of oxygen

---

*This rate also involves a pseudo-steady-state approximation in a mass balance. The oxygen has been treated as in plug flow through a reactor that is uniform in temperature and carbon concentration. If the rate of reaction between carbon and oxygen is first order in oxygen, then it follows from Eq. (4.37) that the overall conversion rate will be the feed flow rate multiplied by

$$1 - \exp(-\text{constant} \times \text{length/velocity}),$$

which is the form of Eq. (5.62). The second term in the denominator of Eq. (5.62b) is the usual kinetic rate term. The first term arises from consideration of a mass transfer limitation on the rate at which oxygen reaches the catalyst surface. The quadratic exponent for R_{ai} is obtained from pilot plant data. This is a much stronger dependence than would normally be expected; a one-half power dependence on flow rate is more common for the mass transfer rate. The combined effects act like two resistances in series, and hence the sum of reciprocals to give the reciprocal of the overall rate.

in the regenerator flue gas; $(21 - O_{fg})/100$ is the oxygen conversion, since the volume fraction in the air feed is 0.21. k_{or} and E_{or} are constants.

Equations (5.55), (5.57), (5.58), (5.60), and (5.61) are a system of five ordinary differential equations for the five variables that define the system state. Some of the process operating variables, such as the air feed rate (R_{ai}) and the catalyst circulation rate (R_{rc}), can be manipulated by an operator or a closed-loop control system in order to change the operating level or to control the process following an upset. Thus, the model, when validated, can be used for design, control, and operating policy decisions.

5.9.3 Validation

Validation was a problem when this model was first developed, and remains a problem to those without access to proprietary FCC operating data. We are relatively comfortable today with the concept that the Kurihara model reflects the important properties of an FCC, because it has been accepted on this basis and used as a frame of reference by oil industry practitioners. In the context of 1967, however, data were not available, except for a report of simulations of an Atlantic Orthoflow-type FCC. Exact operating conditions were not reported, and there were surely major differences in important parameter values. A comparison with the Atlantic results can, however, answer the first essential question in model validation: "Does it *feel* right?"

The Atlantic reactor was controlled by using the flue gas temperature (computed in the model by assuming afterburning of all flue gas oxygen) to adjust the air flow rate, and the reactor temperature to adjust the catalyst recirculation rate. Figures 5.7 and 5.8 show comparisons between the Kurihara model and the reported Atlantic results for the response following a sudden upset in catalyst rate, such as might be caused by a pressure variation; Fig. 5.7 is for a "poor setting" on the temperature controller, and Fig. 5.8 for a "better setting." There are obvious differences in response, but the overall behavior is very close. In particular, the model shows a dominant transient that is of the order of an hour. This is to be compared with catalyst residence times ($H_{ra}/R_{rc}, H_{rg}/R_{rc}$) of the order of a few minutes.

5.9.4 Multiplicity

No detailed study of the steady state characteristics of the Kurihara model has been undertaken. The FCC does have a steady state beyond the normal operating range, corresponding to high regenerator temperature and complete carbon monoxide conversion. The Amoco UltraCat process operates at this state and is claimed to produce a higher gasoline yield than the con-

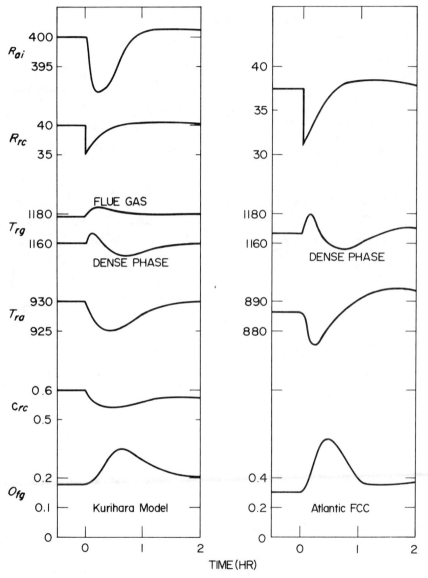

FIG. 5.7.
Dynamical response of Kurihara FCC model and Atlantic reactor following a catalyst rate disturbance for "poor" controller settings (Kurihara, 1967).

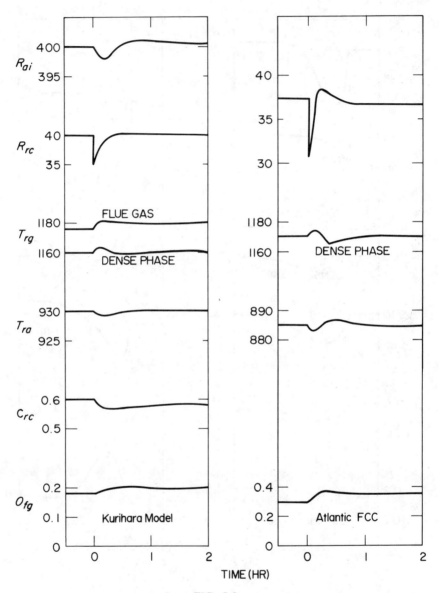

FIG. 5.8.
Dynamical response of Kurihara FCC model and Atlantic reactor following a catalyst rate disturbance for "better" controller settings (Kurihara, 1967).

5.9 FLUID CATALYTIC CRACKER (FCC) MODEL

ventional operating mode. Amoco has apparently developed proprietary technology that enables them to approach this stable state in a safe manner.

Multiplicity has been investigated by Iscol and by Lee and Kugelman, using a model that does not discriminate between carbon types and with simpler kinetic expressions for coke formation and burning. Iscol found multiplicity and potential instability in the normal operating range, while Lee and Kugelman did not. Lee and Weekman attribute the multiplicity in Iscol's results to a cubic temperature dependence in his expression for R_{cf}, while Lee and Kugelman used the same kinetics as Kurihara. This qualitative difference in behavior for a small change in the structure of one constitutive equation points up the importance of studying the sensitivity of the model predictions to structural, as well as parametric, changes.

5.9.5 Order Reduction

Kurihara made use of the difference in residence times of the two reactors to reduce the order and obtain a more tractable set of model equations. (This is the procedure utilized in Section 5.6.) The catalyst residence time in the regenerator is typically 5 min (H_{rg}/R_{rc}, using the typical values in Table 5.1), while the residence time in the cracking reactor is typically 1.5 min (H_{ra}/R_{rc}). Thus, the reactor is expected to respond more quickly, and reactor variables may be considered to be at a steady state on the time scale over which regenerator variables change. On this basis, Kurihara made a pseudo-steady-state approximation with regard to all reactor variables, and set dC_{cat}/dt, dC_{sc}/dt, and dT_{ra}/dt to zero in Eqs. (5.55), (5.57), and (5.58), respectively. The dynamical model is therefore reduced to second order, with Eqs. (5.60) and (5.61) as the model equations. This is a helpful reduction, because second-order dynamical models often lend themselves to graphical analysis. The model also contains the three nonlinear algebraic equations resulting from the pseudo-steady-state approximation, and these do not have a closed-form analytical solution for C_{sc} and T_{ra}.

Figure 5.9 shows a comparison between the dynamical response of the full fifth-order model and the reduced second-order model that includes only regenerator dynamics. It is clear that the important features, particularly the one-hour transient, are retained. There is important process information here that is perhaps obvious in retrospect, but which apparently went unnoticed until publication of the Kurihara model. Operator focus is on the cracking reactor, because it is the unit that is producing the desired product, and reactor output and product variability affect downstream processing units. The control system must focus on the regenerator, because it determines dynamical response. Several authors have discussed this problem from the point of view of operator acceptability and of the impact on downstream processing of control systems designed to speed response.

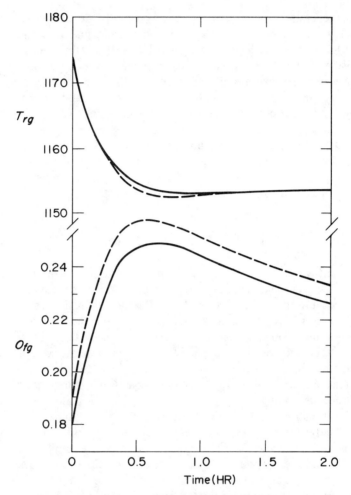

FIG. 5.9.
Open-loop step response of the Kurihara FCC model (solid) and the second-order approximation (dashed) (Isaacs, 1974).

5.9.6 Application

Kurihara's goal was to design a control system that would operate more efficiently than the conventional one. The conventional control scheme measures the reactor temperature and uses this value to adjust catalyst flow rate. It measures flue gas oxygen (in reality, the difference in temperatures between regenerator bed and flue gas, which is a direct measure of O_{fg}), and uses this value to adjust the air rate to the regenerator. Using conventional

5.9 FLUID CATALYTIC CRACKER (FCC) MODEL

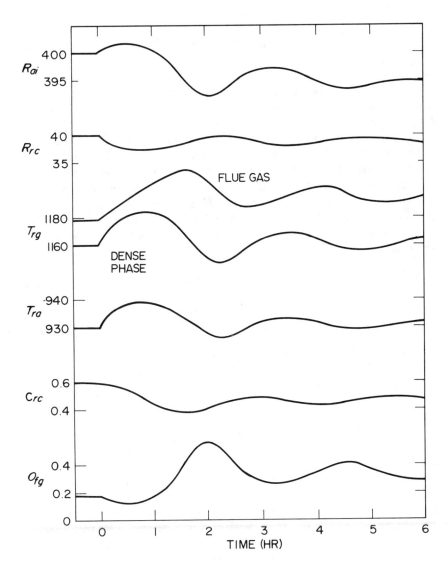

FIG. 5.10.
Response of Kurihara FCC model to 3% decrease in feed rate, conventional control given by Eqs. (5.63). (Kurihara, 1967).

proportional-integral feedback controllers, Kurihara found the best settings for the conventional controller with his model equations to be

$$\Delta R_{rc} = -0.2\Delta T_{ra} - 0.1 \int_0^t \Delta T_{ra}\, dt \qquad (5.63a)$$

$$\Delta R_{ai} = -40\,\Delta O_{fg} - 10\int_0^t \Delta O_{fg}\,dt \qquad (5.63b)$$

The symbol "Δ" denotes the changes in the variable from the set-point value. The response to a 3% decrease in feed rate is shown in Fig. 5.10; note the long transient.

Kurihara exploited the second-order nature of the reduced model to use optimal control theory to design a control system that minimizes an economic performance criterion that he developed, including penalties for violating constraints. The details of this approach are not important, although it is worth noting that the procedure would have been difficult and perhaps even impossible with a model of order higher than second. He ultimately obtained the following practical approximation to his optimal control system:

$$\Delta R_{ai} = -1.0\,\Delta T_{rg} \qquad (5.64a)$$

$$\Delta R_{rc} = +50\,\Delta O_{fg} \qquad (5.64b)$$

The response to the same step change in feed is shown in Fig. 5.11. (Note the expanded scale on the time axis!) The dynamical performance is clearly superior to the conventional scheme.

5.9.7 Final Remarks

The Kurihara scheme does not appear to have been implemented in practice on commercial units, largely because of concerns about the focus on the regenerator. It has reshaped thinking about FCC control. Mobil has patented a modification of the Kurihara scheme that includes a reactor measurement to adjust the set point of the regenerator temperature controller.

We will return to this FCC example for illustration of other aspects of process modeling later in the text. It should be noted again that we have been focusing on only one aspect of FCC control. Detailed modeling of the FCC requires accounting for the many chemical reactions that occur during cracking. It is impossible to keep track of (or even identify) all relevant species, but there has been considerable success in lumping classes of compounds with similar properties into a single pseudo-compound. Component mass balances and rate expressions need only then be written for the lumped variables.

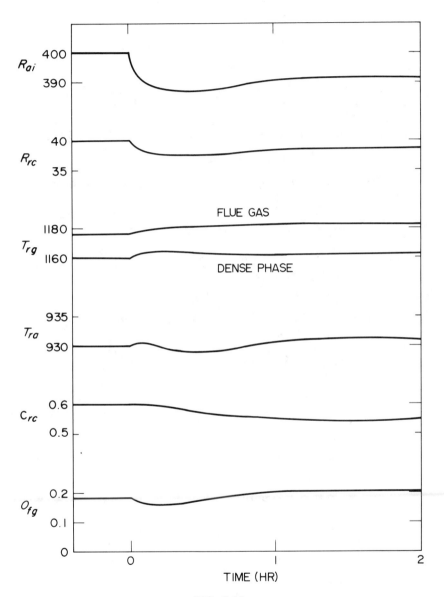

FIG. 5.11.
Response of Kurihara FCC model to 3% decrease in feed rate, approximation to optimal control given by Eqs. (5.64). Note the change in time scale from Fig. 5.10 (Kurihara, 1967).

BIBLIOGRAPHICAL NOTES

The development of the energy balance follows

> T. W. F. Russell and M. M. Denn, *Introduction to Chemical Engineering Analysis* (New York, NY: John Wiley, 1972).
>
> M. M. Denn, "Modeling for Process Control," in C. T. Leondes, ed., *Control and Dynamic Systems: Advances in Theory and Application*, Vol. 15 (New York, NY: Academic Press, 1979).

A similar development is to be found in

> R. Aris, *Introduction to the Analysis of Chemical Reactors* (Englewood Cliffs, NJ: Prentice-Hall, 1965).
>
> R. Aris, *Mathematical Modelling Techniques* (London: Pitman, 1978).

See also Chapter 18 of

> R. B. Bird, W. E. Stewart, and E. N. Lightfoot, *Transport Phenomena* (New York, NY: John Wiley, 1960).

for a general treatment that includes the terms that we have neglected in our derivation. The consequences of ignoring the pV term in Eq. (5.16) for gas phase systems are examined in

> P. M. Mäkilä and K. V. Waller, *Chemical Engineering Science*, **36** (1981) 643.

The way in which an incorrect formulation can be inadvertently introduced with the best intentions is illustrated in the discussion leading to Eq. (9) of

> E. L. Cussler, *Chemical Engineering Education*, **18** (1984) 124.

See also pp. 51ff of

> J. C. Friedly, *Dynamic Behavior of Processes* (Englewood Cliffs, NJ: Prentice-Hall, 1982).

For an example of the commonly used incorrect derivation of the energy equation in an otherwise excellent text, see Sections 2.2 and 8.6 of

> L. A. Gould, *Chemical Process Control: Theory and Applications* (Reading, MA: Addison-Wesley, 1969).

(Our otherwise highly favorable opinion of this important book is expressed in a review in the April, 1973 issue of *IEEE Transactions on Automatic Control*.) An example illustrating the application of the incorrect energy balance can be found in Figs. 9.40 and 9.42 of

> R. G. E. Franks, *Modeling and Simulation in Chemical Engineering* (New York, NY: Wiley-Interscience, 1972).

Application of an incorrect energy balance to a moving bed coal gasifier with steep temperature gradients appears in

> R. Stillman, *IBM J. Res. Dev.*, **23** (1979) 249.
> R. Stillman, *ACS Symposium Series*, **168** (1981) 331.

In a somewhat different mathematical modeling context, the energy balance is derived and reported in the too common incorrect way in a survey of aquatic systems:

> L. T. Fan, R. P. Krishna, and S. H. Lin, "Mathematical Modelling of Aquatic Systems," in A. S. Mujumdar and R. A. Mashelkar, eds., *Advances in Transport Processes*, Vol. II (New Delhi: Wiley Eastern Ltd., 1982, p. 1).

There is a similar error in an otherwise excellent chapter on modeling the process of polymer film blowing in

> C. J. S. Petrie, "Film Blowing, Blow Moulding and Thermoforming," in J. R. A. Pearson and S. M. Richardson, eds., *Computational Analysis of Polymer Processing* (London: Applied Science Publ., 1983, p. 217).

It must be emphasized that this list is simply illustrative, and is by no means complete!

The treatment of multiplicity follows

> M. M. Denn, *Stability of Reaction and Transport Processes* (Englewood Cliffs, NJ: Prentice-Hall, 1975).

See also Aris's modeling book, cited above. A comprehensive study of the possible behavior of Eqs. (5.37) and (5.38) for changing parameter values is given in

> A. Uppal, W. H. Ray, and A. Poore, *Chemical Engineering Science*, **29** (1974) 967; **31** (1976) 205,

where five possible types of steady-state behavior and seventeen types of possible transient behavior are identified. See

> K. F. Jensen and W. H. Ray, *Chemical Engineering Science*, **37** (1982) 199

for an examination of the possible types of behavior in packed tubular reactors, including a tabulation of experimental observations. Schmitz has provided an excellent review of the entire area of multiplicity and dynamical behavior of chemically reacting systems, with extensive references to original research papers and review papers, in

> R. A. Schmitz, *Proc. 1978 Joint Automatic Control Conference*, Vol. II, p. 21 (Pittsburgh: Instrument Soc. of America, 1975).

One of the intriguing references that he cites is

> B. R. Gray, R. Gray, and N. A. Kirwan, *Combustion & Flame*, **118** (1972) 439,

which notes the similarity between ignition and extinction and the accompanying hysterisis (Fig. 5.3) and animal hibernation. For recent experimental and theoretical work on multiplicity and dynamics of catalytic systems, including the relation to the behavior of the stirred tank reactor equations developed here, see

> V. Hlavacek and J. Votruba, *Advances in Catalysis*, **27** (1978) 59.

The comprehensive treatise on the subject is

> R. Aris, *The Mathematical Theory of Diffusion and Reaction in Permeable Catalysts* (Oxford: Clarendon Press, 1975),

especially Vol. 2.

There is a very instructive tutorial introduction to chaotic systems in

> R. M. May, *Nature*, **261** (1976) 459,

where it is shown that the simplest nonlinear difference equation leads to chaos. Another very readable introduction is

> D. Ruelle, *Mathematical Intelligencer*, **2** (1980) 126.

The Lorenz equations are introduced in

> E. N. Lorenz, *J. Atmospheric Sciences*, **20** (1963) 130.

(There is a misprint in Ruelle's article in the second of these equations.) The review by Schmitz contains further references to chaotic behavior in chemically reacting systems. The data in Fig. 5.5 are from

> B. Subramaniam and A. Varma, *Chem. Eng. Communications*, **21** (1983) 221.

The demonstration that chaotic behavior can occur in a continuous-flow stirred tank reactor with only two consecutive reactions is in

> C. Kahlert, O. Rössler, and A. Varma, in "Modeling of Chemical Reaction Systems," *Springer Series in Chemical Physics*, Vol. 18 (New York: Springer Verlag, 1981).

See also

> J. Rathousky, J. Ruszynski, and V. Hlavacek, *Zeit. Naturforsch.*, **35a** (1980) 1238.
>
> J. Rathousky and V. Hlavacek, *J. Chem. Phys.*, **75** (1981) 749.

The first attempt at modeling FCC behavior seems to be

> W. L. Luyben and D. E. Lamb, *Chem. Eng. Progress Symp. Ser. No. 46*, **59** (1963) 165.

The primary source for the Kurihara model is his unpublished dissertation,

> H. Kurihara, *Optimal Control of Fluid Catalytic Cracking Processes*, Sc.D. thesis, M.I.T., Cambridge, MA, 1967, also available as *Report ESL-R-309*, M.I.T. Electronic Systems Laboratory.

The dissertation contains detailed appendixes on the formulation of the rate expressions. The optimal control application is summarized in

> L. A. Gould, L. B. Evans, and H. Kurihara, *Automatica*, **6** (1970) 695,

where references to the data sources for the rate expressions can be found. The parameter values given by Kurihara were modified slightly by Nakano in order to achieve a proper steady state, and the latter's values are reported here. See

> N. Nakano, *Modal Control of a Catalytic Reactor*, M. S. Thesis, M.I.T., Cambridge, MA., 1971.

Kurihara's model has been used in some other theses, where some changes have been made in the parameters:

> M. -T. Eng, *Nonlinear Feedforward Control of the Fluidized Catalytic Cracking Process*, Ph.D. thesis, University of Maryland, College Park, 1973.
>
> B. Isaacs, *Linear-Quadratic Optimal Control of Fluidized Catalytic Cracking Processes*, B.Ch.E. thesis, University of Delaware, Newark, 1974.

Figure 5.9 is from Isaacs. The model and its application have been critically discussed in

> W. Lee and V. W. Weekman, Jr., *AIChE J.*, **22** (1976) 27.
>
> R. Shinnar, in A. S. Foss and M. M. Denn, eds., *Chemical Process Control, AIChE Symposium Series No. 159*, **72** (1977) 167.

The Lee and Weekman paper includes simulations like those in Figs. 5.10 and 5.11 of this text, as well as simulations using the Mobil modification of the Kurihara controller. The FCC models by Iscol and Lee and Kugelman are in

> L. Iscol, *Preprints 1970 Joint Automatic Control Conf.*, St. Louis, MO (1970) 602.
>
> W. Lee and A. M. Kugelman, *Ind. Eng. Chem. Process Des. & Dev.*, **12** (1973) 197.

The lumping required to analyze the product distribution from cracking is described in

> S. M. Jacob, B. Gross, S. E. Voltz, and V. W. Weekman, Jr., *AIChE J.*, **22** (1976) 701.

6

SCALING

6.1 INTRODUCTION

The use of conservation equations to develop a model will usually lead to a mathematical formulation that is very complex. Simplification of the model will be required for analysis and for numerical computation. The first simplification of a model is usually done by estimating the relative importance of the various terms, and eliminating those that are unimportant. This will frequently involve restricting the range of physical application for consistency with the mathematical approximation.

Nondimensionalization is the basic tool for scaling, for it provides the framework for proper order-of-magnitude estimation. We shall illustrate the use of nondimensionalization and order-of-magnitude estimation in this chapter, including several practical examples. Useful results can often be obtained by this approach without the need ever to solve the model equations.

6.2 THE ENERGY BALANCE

It is convenient for what follows to have available the equation describing the temperature distribution in a homogeneous medium. The control volume is shown (in two dimensions, for convenience) in Fig. 6.1. The control volume is of differential size, since the temperature may change spatially. We neglect kinetic and potential energy in the energy equation, which can then be written

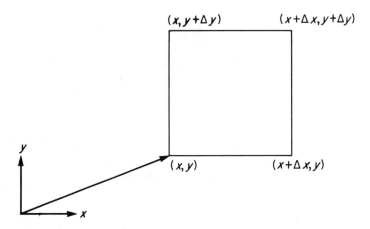

FIG. 6.1
Differential control volume.

$$\frac{\partial}{\partial t}\underline{U}\rho\,\Delta x\,\Delta y\,\Delta z = \rho u\,\Delta y \Delta z \underline{H}|_x - \rho u\,\Delta y \Delta z \underline{H}|_{x+\Delta x}$$

$$+ \rho v\,\Delta x\,\Delta z \underline{H}|_y - \rho v\,\Delta x\,\Delta z \underline{H}|_{y+\Delta y}$$

$$+ \rho w\,\Delta x\,\Delta y \underline{H}|_z - \rho w\,\Delta x\,\Delta y \underline{H}|_{z+\Delta z}$$

$$+ q_x\,\Delta y\,\Delta z|_x - q_x\,\Delta y\,\Delta z|_{x+\Delta x}$$

$$+ q_y\,\Delta x\,\Delta z|_y - q_y\,\Delta x\,\Delta z|_{y+\Delta y}$$

$$+ q_z\,\Delta x\,\Delta y|_z - q_z\,\Delta x\,\Delta y|_{z+\Delta z}$$

Here, u, v, and w represent the x-, y-, and z- components, respectively, of the velocity vector, **v**. Thus, $\rho u\,\Delta y\,\Delta z$ is the mass flow rate across the control volume face that is orthogonal to the x-direction. The work required to put material into and out of the control volume has been included, which is the reason that the energy flow terms are written in terms of enthalpy per unit mass, $\underline{H}$, rather than internal energy, $\underline{U}$. q_x is the rate of heat flow per unit surface area in the x-direction; since heat flow is directional, it can be represented by a vector, **q**, of which q_x is the x-component. The time derivative is a partial derivative because it is applied at the particular spatial position where the control volume is located. The terms which account for work done on the control volume by the viscous stresses in the surrounding material have been neglected; one consequence is that terms that can be identified with viscous dissipation do not appear.

If we divide each term in Eq. (6.1) by $\Delta x\,\Delta y\,\Delta z$ and take the limit

6.2 THE ENERGY BALANCE

as $\Delta x \to 0$, $\Delta y \to 0$, $\Delta z \to 0$, we obtain the partial differential equation*

$$\frac{\partial}{\partial t}\rho \underline{U} = -\frac{\partial}{\partial x}\rho u \underline{H} - \frac{\partial}{\partial y}\rho v \underline{H} - \frac{\partial}{\partial z}\rho w \underline{H}$$

$$-\frac{\partial q_x}{\partial x} - \frac{\partial q_y}{\partial y} - \frac{\partial q_z}{\partial z}$$

$$= -\nabla \cdot \rho \mathbf{v} \underline{H} - \nabla \cdot \mathbf{q} \qquad (6.2)$$

By replacing $\underline{U}$ with $\underline{H} - p/\rho$, and neglecting the term $\partial p/\partial t$ (valid for liquids and systems at constant pressure), we obtain the equation

$$\rho \frac{\partial \underline{H}}{\partial t} = -\rho \mathbf{v} \cdot \nabla \underline{H} - \nabla \cdot \mathbf{q} \qquad (6.3)$$

Here, we have also utilized the equation of conservation of mass (known as the *continuity equation*),

$$\frac{\partial \rho}{\partial t} = -\nabla \cdot \rho \mathbf{v} \qquad (6.4)$$

which is derived in the same way. Finally, if the composition is uniform in space and time, and the pressure dependence of the enthalpy can be neglected, then $\underline{H}$ is a function only of T and

$$\frac{\partial \underline{H}}{\partial t} = c_p \frac{\partial T}{\partial t}, \qquad \nabla \underline{H} = c_p \nabla T$$

---

*∇ is the gradient operator,

$$\mathbf{i}_x \frac{\partial}{\partial x} + \mathbf{i}_y \frac{\partial}{\partial y} + \mathbf{i}_z \frac{\partial}{\partial z},$$

where $\mathbf{i}_x$, $\mathbf{i}_y$, $\mathbf{i}_z$ are the unit vectors in the x-, y-, and z-coordinate directions, respectively. The operator $\nabla \cdot$ operating on a vector $\mathbf{A}$ is a scalar, $\partial A_x/\partial x + \partial A_y/\partial y + \partial A_z/\partial z$ in Cartesian coordinates. ∇A, with A a scalar, is the vector

$$\mathbf{i}_x \frac{\partial A}{\partial x} + \mathbf{i}_y \frac{\partial A}{\partial y} + \mathbf{i}_z \frac{\partial A}{\partial z}.$$

The operator $\nabla \cdot \nabla$, or ∇^2, is the Laplacian, $\partial^2/\partial x^2 + \partial^2/\partial y^2 + \partial^2/\partial z^2$ in Cartesian coordinates.

Thus,

$$\rho c_p \frac{\partial T}{\partial t} + \rho c_p \mathbf{v} \cdot \nabla T = - \nabla \cdot \mathbf{q} \tag{6.5}$$

We require a constitutive equation for the heat flux vector, $\mathbf{q}$. Fourier's law, which states that $\mathbf{q}$ is colinear with ∇T, describes most data for homogeneous isotropic systems:

$$\mathbf{q} = - k \nabla T \tag{6.6}$$

k is known as the *thermal conductivity*, and might depend on temperature and composition. Thus, the final form of the energy equation is

$$\rho c_p \frac{\partial T}{\partial t} + \rho c_p \mathbf{v} \cdot \nabla T = \nabla \cdot (k \nabla T) \tag{6.7}$$

If k is spatially constant, then we have the more commonly used form,

$$\rho c_p \frac{DT}{Dt} = \rho c_p \frac{\partial T}{\partial t} + \rho c_p \mathbf{v} \cdot \nabla T = k \nabla^2 T \tag{6.8}$$

The operator $D/Dt = \partial/\partial t + \mathbf{v} \cdot \nabla$ is known as the *substantial derivative*; it describes the rate of change with time in a frame of reference that moves with the material.

Equation (6.7) requires boundary conditions. These are most easily introduced by considering a planar boundary on the x-axis through which no mass flows. The only mechanism available for energy to be transferred through the boundary is the flow of heat: q_y equals $-k\, \partial T/\partial y$. The "driving force" for heat transfer will be the difference between the temperature near the boundary in the material described by Eq. (6.7) and the temperature of the bounding medium, which we will denote by T_∞. It is conventional to define a *heat transfer coefficient*, denoted h, as

$$h = q_y/(T - T_\infty) \tag{6.9}$$

h depends on all the properties of the system, and a proper constitutive equation is required. We shall discuss this subsequently, where we will see how an ordering analysis can at times provide the constitutive equation directly. Equation (6.9) is sometimes called "Newton's law of cooling." It is in reality nothing more than a definition of h, since q_y and $T - T_\infty$ can be measured directly.

6.3 THE HEAT EQUATION

Now, let us consider the case of a homogeneous quiescent (solid or liquid) slab of thickness $2L$ in the y-direction and extending infinitely far in the x- and z-directions. The slab is initially at a uniform temperature, T_0. At time $t = 0$ it is suddenly exposed to a constant temperature T_∞ on the faces $y = \pm L$. For use in our later discussion of scaling, we wish to compute the temperature as a function of y and t.

We assume uniformity in the x- and z-directions, and that $k =$ constant. In the absence of the flow, Eq. (6.7) then becomes

$$\rho c_p \frac{\partial T}{\partial t} = k \frac{\partial^2 T}{\partial y^2} \tag{6.10}$$

Equation (6.10) is often called the *heat equation* (and sometimes the *diffusion equation*, since the same equation arises in mass transfer). We assume symmetry about the center plane, $y = 0$, which gives one boundary condition,

$$\frac{\partial T}{\partial y} = 0 \quad \text{at} \quad y = 0 \tag{6.11a}$$

For the other boundary conditions we assume that we know the heat transfer coefficient at the surface, in which case Eq. (6.9) becomes

$$-k \frac{\partial T}{\partial y} = h(T - T_\infty) \quad \text{at} \quad y = L \tag{6.11b}$$

We are also given the initial condition,

$$T = T_0 \quad \text{for all } y \text{ at } t = 0 \tag{6.11c}$$

Most people would probably choose to solve Eq. (6.10) using the method of separation of variables. This method can be used only when the boundary conditions are *homogeneous;* that is, a linear combination of the dependent variable and its derivatives must sum to zero at the boundary. Equation (6.11b) is not homogeneous, and the dependent variable must be changed from T to $T - T_\infty$. Since the dependent variable will then go to zero as $t \to \infty$, it is common to divide by $T_0 - T_\infty$ in order to make the initial condition unity. Thus, defining

$$\hat{T} = \frac{T - T_\infty}{T_0 - T_\infty} \tag{6.12}$$

Eqs. (6.10) and (6.11) become

$$\rho c_p \frac{\partial \hat{T}}{\partial t} = k \frac{\partial^2 \hat{T}}{\partial y^2} \tag{6.13}$$

$$\frac{\partial \hat{T}}{\partial y} = 0 \quad \text{at} \quad y = 0 \tag{6.14a}$$

$$-k \frac{\partial \hat{T}}{\partial y} = h\hat{T} \quad \text{at} \quad y = L \tag{6.14b}$$

$$\hat{T} = 1 \quad \text{at} \quad t = 0 \tag{6.14c}$$

The solution to Eq. (6.13) with the boundary conditions (6.14a,b), for an arbitrary symmetric initial temperature distribution, is

$$\hat{T}(y,t) = \sum_{n=0}^{\infty} C_n \exp(-\lambda_n^2 kt/\rho c_p L^2) \cos(\lambda_n y/L) \tag{6.15}$$

The coefficients $\{C_n\}$ are determined from the initial condition and do not really interest us here. The *eigenvalues* $\{\lambda_n\}$ are solutions of the transcendental equation

$$\lambda_n \tan \lambda_n = hL/k \equiv \text{Bi}; \quad n = 0, 1, 2, \ldots, \infty \tag{6.16}$$

where Bi is a dimensionless heat transfer coefficient known as the *Biot number*.* The solutions to Eq. (6.16) for $n \geq 1$ are relatively insensitive to Bi and are bounded by

$$n\pi \leq \lambda_n \leq (n + \tfrac{1}{2})\pi; \quad n = 1, 2, \ldots, \infty \tag{6.17}$$

---

*The Biot number has a physical interpretation. At steady state, the rate of conduction to the wall in the medium will be roughly $k[T(0) - T(L)]/L$, while the rate of heat transfer to the surroundings is $h[T(L) - T_\infty]$. These rates must be equal, so the Biot number is a measure of the ratio of internal to external temperature differences: $\text{Bi} \sim [T(0) - T(L)]/[T(L) - T_\infty]$. Bi will be large when the important resistance is in the medium and not the surroundings $[T(L) \approx T_\infty]$, and this is the only case in which we will be interested. The Biot number for the moving bed coal gasifier is about 100, to give an idea of the expected magnitude. A similar dimensionless group, in which k refers to the thermal conductivity of the *surroundings*, is known as the *Nussult number*.

6.4 SCALING THE HEAT EQUATION

The lower bound corresponds to Bi → 0 (nearly adiabatic surface), while the upper bound corresponds to Bi → ∞ (no resistance to heat transfer at the surface). The first eigenvalue lies between zero and $\pi/2$; for Bi = 1, $\lambda_0 = 0.86$, while for small Bi, $\lambda_0 \sim \text{Bi}^{1/2}$.

The arguments of successive exponentials increase rapidly, so the series in Eq. (6.16) will usually be dominated by the first term (corresponding to the smallest value of λ_n). For Bi at least of order unity, the smallest eigenvalue will be between about one and $\pi/2$, so the solution to Eq. (6.15) is dominated by a term

$$\hat{T} \sim \exp(-\pi^2 kt/4\rho c_p L^2) \tag{6.18}$$

The steady state is essentially reached when the argument of the exponential is between -2 to -3, depending on the degree of approach required. Since $\pi^2/4 \sim 2.5$, we can say that steady state will essentially be reached for $t \sim \rho c_p L^2/k$. That is, *there is a characteristic quantity that defines the time scale of the transient*. The transient nature of the process must be taken into account with respect to transient events that occur on this time scale. The process responds as though it were in steady state to events that occur on a much slower time scale, while the process cannot respond at all to events that occur on a much faster time scale.

We can be somewhat more specific about this last statement, which may not be obvious. Suppose the temperature external to the slab changes very slowly when time is measured in units of $\rho c_p L^2/k$. In that case, the temperature everywhere in the slab will be close to the external temperature, with a time lag that is only of order $\rho c_p L^2/k$ and thus not important with respect to the time scale of process changes. If the external temperature is changed over a time that is very rapid relative to $\rho c_p L^2/k$, however, the temperature change is not observed internally until a time that is $\rho c_p L^2/k$ later. Thus, the system will not respond at all to temperature oscillations with a frequency that is fast relative to $k/\rho c_p L^2$.

There is an alternative interpretation that is often useful, and will be employed in the modeling of the moving bed coal gasifier. If we change a temperature at the boundary, the effect of this change after a time t will have penetrated a distance roughly equal to $L = (kt/\rho c_p)^{1/2}$ into the interior.

6.4 SCALING THE HEAT EQUATION

The conclusions that we reached regarding the heat equation at the end of the preceding section are clearly important with regard to understanding the behavior that we might expect. It was necessary to obtain a solution to the partial differential equation, however, in order to do so. We now show how the same conclusions can be reached by scaling the equation.

The basic idea in scaling is to estimate the magnitude of each dependent and independent variable, and then to define new (usually dimensionless) variables with a range from zero to unity. This is done by using the estimated magnitudes as the measurement scales. If we do the scaling properly, then each term in the equation will be of order unity, possibly multiplied by a dimensionless group containing system parameters. This formulation will enable us to estimate the importance of each term.

The temperature ranges from T_∞ outside the boundary to the maximum or minimum within the slab, so $T - T_\infty$ is the appropriate dependent variable. The maximum for this quantity is $T_0 - T_\infty$, so the dimensionless dependent variable $\hat{T}$ defined by Eq. (6.12) does range from zero to a maximum value of roughly unity. The length scale over which changes in temperature occur is L, so the appropriate dimensionless length coordinate is $\hat{y} = y/L$. The maximum change in $\hat{T}$ is of order unity, and it occurs over a dimensionless distance that is of order unity; thus, dimensionless spatial *derivatives* of temperature are expected to be of order unity.

The time scale is not obvious (or would not be obvious, had we not solved the equation previously). We will call the time scale θ, and write the dimensionless time as $\hat{t} = t/\theta$. The characteristic time is a measure of the time over which $\hat{T}$ changes by order unity, so dimensionless time derivatives will be of order unity (change in dimensionless temperature of order unity/unit change in dimensionless time). Equation (6.9) therefore becomes*

$$\frac{\rho c_p}{\theta}\frac{\partial \hat{T}}{\partial \hat{t}} = \frac{k}{L^2}\frac{\partial^2 \hat{T}}{\partial \hat{y}^2} \tag{6.19}$$

or

$$\left(\frac{\rho c_p L^2}{k\theta}\right)\frac{\partial \hat{T}}{\partial \hat{t}} = \frac{\partial^2 \hat{T}}{\partial \hat{y}^2} \tag{6.20}$$

The dimensionless derivatives are of order unity. The transient term (the left-hand side) will be comparable to the spatial distribution term (the right-hand side) only if the coefficient of $\partial \hat{T}/\partial \hat{t}$ is of order unity; i.e.,

$$\theta = \rho c_p L^2 / k \tag{6.21}$$

Thus, this rather simple ordering analysis establishes the natural time scale

---

*$\partial^2 T = (T_0 - T_\infty)\, \partial^2 \hat{T}$; $\partial t = \partial(\theta \hat{t}) = \theta\, \partial \hat{t}$; $\partial y^2 = \partial(L\hat{y})^2 = L^2 \partial \hat{y}^2$.

of the system. The discussion at the end of the preceding section then follows, but the conclusions regarding "fast" and "slow" processes can be reached without any need to solve the partial differential equation.

The time scale in Eq. (6.21) is sometimes called the *Einstein diffusion time*, since it arises naturally in Einstein's analysis of Brownian motion as a random walk process. Heat conduction, mass diffusion, and viscous stress transmission are all Brownian processes at the molecular level. θ is also sometimes called the *Fourier time*, since it arises in Fourier's analysis of transient conduction. We can, equivalently, define the *Einstein length*, $L = (kt/\rho c_p)^{1/2}$, as the length scale over which changes at the boundary are felt over a time t.

6.5 APPLYING THE EINSTEIN TIME/LENGTH CONCEPT

We can illustrate the usefulness of scaling the heat equation by two examples, each of which we will only sketch. The first is a real industrial problem that was analyzed in some detail, when an ordering argument would have sufficed to show that the analysis was mostly unnecessary. The application is in fiber spinning, described in Section 2.5, where there is sometimes a "cold drawing" step in which the solidified polymer undergoes a rapid area reduction. The corresponding change in structure is exothermic, and it is important to know how much of the enthalpy change goes into increasing the temperature (and perhaps melting crystals), and how much is lost to the surrounding air.

Typical polymer properties in cgs units are $\rho \sim 1$, $c_p \sim 0.3$, and $k \sim 5 \times 10^{-4}$. Thus, $\rho c_p/k \sim 10^3$ s/cm^2. A typical filament radius is 10^{-2} cm, so $\theta \sim 0.1$ sec. The rapid area reduction takes place over a millisecond time scale, in which case there is effectively no heat transfer between the interior of the filament and the surroundings. The cold draw process can therefore be considered to be adiabatic, and all of the enthalpy change associated with the change in structure must go into a temperature increase.

The moving bed coal gasifier was described in Section 2.4. We noted that the gasifier is cooled at the wall. Thus, we expect radial conduction of heat, and hence radial temperature gradients. Our description of the reactor will therefore require that we account for axial and radial temperature variations, leading to a model in terms of partial differential equations even to describe the steady state. We will, in fact, ultimately utilize such a description of the reactor, but this will be at the "fine-tuning" stage and only after we have gained much useful process information from a model that requires only ordinary differential equations.

If we follow a coal particle on its path through the reactor, the particle

112   6   SCALING

is an element of a slab (actually a cylinder) that has been in contact with a new environmental temperature for a time equal to the time in the reactor. The residence time for a coal particle from top to bottom of a Lurgi dry ash reactor operating at the normal throughput is about 50 min, or 3000 sec. The density of coal in cgs units is about 1.3, and the heat capacity is about 0.3. The thermal conductivity is about 5×10^{-4}, but gas mixing in the interstices and particle-to-particle radiation raise the effective conductivity to about 6×10^{-3}. Thus, the Einstein length at the end of the period of travel is

$$L \sim \left[\frac{(6 \times 10^{-3}) \times (3 \times 10^{+3})}{1.3 \times 0.3} \right]^{1/2} \sim 7 \text{ cm}^*$$

The reactor radius is typically from 1.5 to 2 m, so the wall cooling affects only a small part of the cross section. Hence, to a good approximation, the reactor can be treated as though the radial temperature profile were uniform, as in an adiabatic reactor, with perhaps some special attention paid separately to the small region near the wall in order to estimate the rate of heat loss to the jacket and the steam consumption. The model that can be employed to describe steady-state operation therefore consists of ordinary differential equations.

This simple calculation leads to a substantial simplification in computational effort. We can also see that the wall region will comprise a major portion of the reactor cross section when the transit time has decreased by a factor of 100, as would happen when the throughput has been turned down by a large amount. In that case, a complete treatment of radial effects will be required.

6.6   TRANSPORT OF MASS, MOMENTUM, AND ENERGY

6.6.1   Problem Statement

The power of scaling for model simplification, and for reaching conclusions about system performance without computation, is nicely illustrated by considering the problem of estimating the heat transfer coefficient to a fluid at a solid surface. The equations for conservation of mass, momentum, and energy for two-dimensional flow of an incompressible fluid without chemical reaction in a rectangular cartesian coordinate system are summarized in Table 6.1. Equation (V) corresponds to Eq. (6.7), but for completeness it

---

*A similar conclusion is reached if gas properties are used. The gas transit time is about one minute.

6.6 TRANSPORT OF MASS, MOMENTUM, AND ENERGY

includes the viscous dissipation terms. The mass and momentum equations are derived in a similar way to the derivation of Eq. (6.7), using the same differential control volume. The pressure includes the gravitational contribution. The constitutive equation for a Newtonian fluid is given by Eqs. (III); η is the shear viscosity, which is a material parameter. η may be temperature dependent and even pressure dependent at very high pressures, but it is taken in this section to be constant. Substitution of Eqs. (III) into the momentum conservation equation (II) with constant η leads to the *Navier-Stokes* equations (IV).

This set of nonlinear partial differential equations is extremely complex, and few exact solutions are available even for isothermal flows. Thus, simplification by means of scaling is essential if complex problems are to be analyzed. We will give one example here of the large amount of information that can be deduced in the course of simplification and prior to any actual solution. The problem is shown schematically in Fig. 6.2. An incompressible Newtonian fluid flows at high speed past a sharp-edged plate. The fluid velocity far from the plate is parallel to the plate surface and has a magnitude U. The plate is maintained at temperature T_w, while the fluid in the "free stream" is at T_∞. The flow is steady ($\partial/\partial t = 0$). We wish to know the drag on the plate and the rate of heat transfer between the plate and the fluid.

6.6.2 Fluid Flow

We need first to deal with the description of the motion. The velocity parallel to the plate goes from a value of zero at the surface (the *no-slip condition*, which is a cornerstone of classical hydrodynamics) to a value of U at some distance from the surface. Thus, U is the appropriate scaling parameter, and the dimensionless x-component of velocity will be $\hat{u} = u/U$. If the plate is characterized by some length L, then changes in the x-direction will take place over that distance, and the dimensionless x-coordinate will be taken as $\hat{x} = x/L$.

Changes in the y-direction will occur over a small region (known as the *boundary layer*) between the uniform external flow and the no-slip surface.* This distance will change somewhat with position, but we expect it always to be small. We will let δ denote this boundary layer scale for x close to L, and we note that we expect $\delta/L \ll 1$. The actual magnitude of δ will be one of the outcomes of our ordering analysis. We therefore write $\hat{y} = y/\delta$. Finally, the bulk of the flow is parallel to the surface, but the presence of the plate will cause a small amount of flow normal to the plate.

---

*This analysis assumes at least a basic introduction to fluid flow, and rather standard arguments are being employed. The reader to whom the Navier-Stokes equations are totally new may wish to review some elementary fluid mechanics before continuing.

TABLE 6.1
Equations of mass, momentum, and energy conservation for two-dimensional flow of an incompressible fluid

Mass (incompressible): $\dfrac{\partial u}{\partial x} + \dfrac{\partial v}{\partial y} = 0$ \hfill (I)

Momentum:

x-component: $\rho\left(\dfrac{\partial u}{\partial t} + u\dfrac{\partial u}{\partial x} + v\dfrac{\partial u}{\partial y}\right) = -\dfrac{\partial p}{\partial x} + \dfrac{\partial \tau_{xx}}{\partial x} + \dfrac{\partial \tau_{yx}}{\partial y}$ \hfill (IIa)

y-component: $\rho\left(\dfrac{\partial v}{\partial t} + u\dfrac{\partial v}{\partial x} + v\dfrac{\partial v}{\partial y}\right) = -\dfrac{\partial p}{\partial y} + \dfrac{\partial \tau_{xy}}{\partial x} + \dfrac{\partial \tau_{yy}}{\partial y}$ \hfill (IIb)

Constitutive equation (Newtonian fluid): $\tau_{xx} = 2\eta\dfrac{\partial u}{\partial x}, \quad \tau_{yy} = 2\eta\dfrac{\partial v}{\partial y}$ \hfill (IIIa,b)

$$\tau_{xy} = \tau_{yx} = \eta\left(\dfrac{\partial u}{\partial y} + \dfrac{\partial v}{\partial x}\right)$$ \hfill (IIIc)

Navier-Stokes equations (Newtonian fluid):

x-component: $\rho\left(\dfrac{\partial u}{\partial t} + u\dfrac{\partial u}{\partial x} + v\dfrac{\partial u}{\partial y}\right) = -\dfrac{\partial p}{\partial x} + \eta\left(\dfrac{\partial^2 u}{\partial x^2} + \dfrac{\partial^2 u}{\partial y^2}\right)$ \hfill (IVa)

y-component: $\rho\left(\dfrac{\partial v}{\partial t} + u\dfrac{\partial v}{\partial x} + v\dfrac{\partial v}{\partial y}\right) = -\dfrac{\partial p}{\partial y} + \eta\left(\dfrac{\partial^2 v}{\partial x^2} + \dfrac{\partial^2 v}{\partial y^2}\right)$ \hfill (IVb)

Energy (constant thermal conductivity, incompressible Newtonian fluid):

$$\rho c_p\left(\dfrac{\partial T}{\partial t} + u\dfrac{\partial T}{\partial x} + v\dfrac{\partial T}{\partial y}\right) = k\left(\dfrac{\partial^2 T}{\partial x^2} + \dfrac{\partial^2 T}{\partial y^2}\right) + \eta\left[2\left\{\left(\dfrac{\partial u}{\partial x}\right)^2 + \left(\dfrac{\partial v}{\partial y}\right)^2\right\} + \left(\dfrac{\partial v}{\partial x} + \dfrac{\partial u}{\partial y}\right)^2\right]$$ \hfill (V)

6.6 TRANSPORT OF MASS, MOMENTUM, AND ENERGY

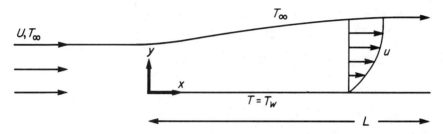

FIG. 6.2
Schematic of laminar flow over a flat plate with a sharp leading edge.

We will denote the magnitude of this flow in the y-direction by V, and take the dimensionless y-component of velocity as $\hat{v} = v/V$. V is also unknown at this point.

The continuity equation (I) is written in terms of the dimensionless variables as

$$\frac{U}{L}\frac{\partial \hat{u}}{\partial \hat{x}} + \frac{V}{\delta}\frac{\partial \hat{v}}{\partial \hat{y}} = 0 \qquad (6.22a)$$

or

$$\left(\frac{U\delta}{VL}\right)\frac{\partial \hat{u}}{\partial \hat{x}} + \frac{\partial \hat{v}}{\partial \hat{y}} = 0 \qquad (6.22b)$$

The dimensionless derivatives are of order unity (dimensionless changes of order unity over dimensionless distances of order unity) *if* we have chosen the scaling variables correctly. The two terms in the dimensionless continuity equation must be comparable, for otherwise a contradiction results: if the first term is much larger than the second, then we can neglect the second relative to the first, in which case the first stands alone equal to zero, and similarly if the second is much larger than the first. Since the derivatives are of order unity, the terms are of comparable magnitude only if the dimensionless grouping $U\delta/VL$ is of order unity. Thus, the magnitude of V is established in terms of δ:

$$V = U\delta/L \qquad (6.23)$$

As expected, $V \ll U$.

We now turn to the x-component of the Navier-Stokes equations. This equation also involves the pressure, which must be scaled. From a basic knowledge of hydrodynamics, we expect the pressure in a high-speed flow to be of order ρU^2, but we can let this result come out of the analysis. We will call the characteristic pressure Π, and define the dimensionless pressure as $\hat{p} = p/\Pi$. Equation (IVa), with $\partial/\partial t = 0$, is then written in dimensionless form as

$$\frac{\rho U^2}{L}\left(\hat{u}\frac{\partial \hat{u}}{\partial \hat{x}} + \hat{v}\frac{\partial \hat{u}}{\partial \hat{y}}\right) = -\frac{\Pi}{L}\frac{\partial \hat{p}}{\partial \hat{x}} + \frac{\eta U}{\delta^2}\left[\left(\frac{\delta}{L}\right)^2 \frac{\partial^2 \hat{u}}{\partial \hat{x}^2} + \frac{\partial^2 \hat{u}}{\partial \hat{y}^2}\right] \quad (6.24)$$

Here, we have already made use of Eq. (6.23) and replaced V with $U\delta/L$.

One simplification follows immediately: $\partial^2 \hat{u}/\partial \hat{x}^2$ and $\partial^2 \hat{u}/\partial \hat{y}^2$ are both of order unity (assuming correct scaling), but the former is multiplied by $(\delta/L)^2$, which is very small. Thus, we can neglect the x-derivative term relative to the y-derivative. This is an example of simplification through scaling. With this simplification we rearrange Eq. (6.24) slightly to obtain

$$\left(\frac{\rho U \delta^2}{\eta L}\right)\left(\hat{u}\frac{\partial \hat{u}}{\partial \hat{x}} + \hat{v}\frac{\partial \hat{u}}{\partial \hat{y}}\right) = -\left(\frac{\Pi \delta^2}{\eta U L}\right)\frac{\partial \hat{p}}{\partial \hat{x}} + \frac{\partial^2 \hat{u}}{\partial \hat{y}^2} \quad (6.25)$$

The terms on the left are the inertial terms, which must be important in a high-speed flow. The far right term is the shear stress, which must be important near the wall. Thus, these terms must be comparable. Since the scaled derivative terms are all of order unity, it follows that $(\rho U \delta^2/\eta L)$ must also be of order unity, thus establishing the scaling factor in the y-direction:

$$\delta = (\eta L/\rho U)^{1/2} = L\,\mathrm{Re}_L^{-1/2} \quad (6.26)$$

Re is the symbol for the *Reynolds number*. The symbol Re_x is often used in boundary layer theory to denote a Reynolds number based on distance from the leading edge:

$$\mathrm{Re}_x = xU\rho/\eta \quad (6.27)$$

Thus, $\mathrm{Re}_L = LU\rho/\eta$. It is clear that on a long flat plate which does not have a well-defined length (an "infinite" plate), L can be taken to be any distance downstream of the leading edge, so it follows from Eq. (6.26) that δ is in fact a function of distance from the leading edge.

The pressure gradient term is also expected in a high-speed flow to be comparable to the inertial terms, so $\Pi\delta^2/\eta UL$ must also be of order unity. This fixes the characteristic pressure:

6.6 TRANSPORT OF MASS, MOMENTUM, AND ENERGY

$$\Pi = \frac{\eta U L}{\delta^2} = \rho U^2 \tag{6.28}$$

This is the expected result.

We now turn to the y-component of the Navier-Stokes equations, (IVb). After writing this equation in dimensionless form, using Eqs. (6.23) and (6.28) for V and Π, we can rearrange the equation to the form

$$\frac{\partial \hat{p}}{\partial \hat{y}} = \left(\frac{\delta}{L}\right)^2 \left[\left(\frac{\delta}{L}\right)^2 \frac{\partial^2 \hat{v}}{\partial \hat{x}^2} + \frac{\partial^2 \hat{v}}{\partial \hat{y}^2} - \hat{u}\frac{\partial \hat{v}}{\partial \hat{x}} - \hat{v}\frac{\partial \hat{v}}{\partial \hat{y}}\right] \tag{6.29}$$

The term in brackets is of order unity, and it is multiplied by $(\delta/L)^2$. Thus, $\partial \hat{p}/\partial \hat{y}$ is of order $(\delta/L)^2$, so $\hat{p}$ is independent of $\hat{y}$ to within the approximation that we have been using. The y-component therefore simplifies to $\hat{p} = \hat{p}(\hat{x})$, which is a major simplification. We can therefore summarize the *boundary layer equations* in dimensional form as

$$p = p(x) \tag{6.30a}$$

$$\rho \left(u\frac{\partial u}{\partial x} + v\frac{\partial u}{\partial y}\right) = -\frac{dp}{dx} + \eta \frac{\partial^2 u}{\partial y^2} \tag{6.30b}$$

These equations are much simpler than Eqs. (IV), and solutions do exist for problems of interest, including the flat plate.

The drag on the plate is usually expressed in terms of a *drag coefficient*,

$$C_D = \frac{\tau_{xy}}{\frac{1}{2}\rho U^2} = \frac{\eta}{\frac{1}{2}\rho U^2}\left(\frac{\partial u}{\partial y} + \frac{\partial v}{\partial x}\right) = \frac{\eta}{\frac{1}{2}\rho U \delta}\left[\frac{\partial \hat{u}}{\partial \hat{y}} + \left(\frac{\delta}{L}\right)^2 \frac{\partial \hat{v}}{\partial \hat{x}}\right] \tag{6.31}$$

The second term in the square bracket is negligible, while the first is of order unity. Thus, using Eq. (6.26) for δ, and dropping the factor of $\frac{1}{2}$ for an ordering estimate, we have

$$C_D \sim \frac{\eta}{\rho U x \, \text{Re}_x^{-1/2}} \sim \text{Re}_x^{-1/2} \tag{6.32}$$

Here, we have used x in place of L to account for the fact that L could represent any position away from the leading edge. The complete solution to Eqs. (6.30) gives the coefficient in Eq. (6.32),

$$C_D = 0.664 \, \text{Re}_x^{-1/2} \qquad (6.33)$$

but this result requires solution of a nonlinear partial differential equation.

6.6.3 Heat transfer

We can now turn to the heat transfer problem. As before, it is only temperature differences that are of concern, so the dimensionless variable that is appropriate is

$$\hat{T} = \frac{T - T_w}{T_\infty - T_w} \qquad (6.34)$$

We expect the "thermal boundary layer" over which the temperature changes from the wall to the free stream value to be roughly the same as δ, but there might be some differences, and we will denote the y-direction scaling length by δ_T. We will retain the symbol $\hat{y}$ for y/δ, and use $\hat{y}_T$ for y/δ_T. The dimensionless form of the steady-state energy equation is therefore (after slight rearrangement)

$$\left(\frac{\rho c_p U \delta_T^2}{kL}\right) \left[\hat{u}\frac{\partial \hat{T}}{\partial \hat{x}} + \left(\frac{\delta}{\delta_T}\right) \hat{v}\frac{\partial \hat{T}}{\partial \hat{y}_T}\right]$$

$$= \left[\left(\frac{\delta_T}{L}\right)^2 \frac{\partial^2 \hat{T}}{\partial \hat{x}^2} + \frac{\partial^2 \hat{T}}{\partial \hat{y}_T^2}\right] + \frac{\eta U^2}{k(T_\infty - T_w)} \left(\frac{\delta_T}{\delta}\right)^2$$

$$\times \left\{\left[\frac{\partial \hat{u}}{\partial \hat{y}} + \left(\frac{\delta}{L}\right)^2 \frac{\partial \hat{v}}{\partial \hat{x}}\right]^2 + \left(\frac{\delta}{L}\right)^2 \left[\left(\frac{\partial \hat{u}}{\partial \hat{x}}\right)^2 + \left(\frac{\partial \hat{v}}{\partial \hat{y}}\right)^2\right]\right\} \qquad (6.35)$$

One immediate simplification is that the $\partial^2 \hat{T}/\partial \hat{x}^2$ term can be dropped, since $(\delta_T/L)^2 \ll 1$, and only the $\partial \hat{u}/\partial \hat{y}$ term remains in the final (dissipation) term. The *Griffith number*, $\eta U^2/k(T_\infty - T_w)$, will often be small, in which case the dissipation term can be neglected relative to the conduction term. We shall assume a small Griffith number here for convenience. In that case, the working form of the energy equation is

$$\frac{\rho c_p U \delta_T^2}{kL} \left[\hat{u}\frac{\partial \hat{T}}{\partial \hat{x}} + \left(\frac{\delta}{\delta_T}\right) \hat{v}\frac{\partial \hat{T}}{\partial \hat{y}_T}\right] = \frac{\partial^2 \hat{T}}{\partial \hat{y}_T^2} \qquad (6.36)$$

The ordering analysis from this point is a bit delicate, since we are

6.6 TRANSPORT OF MASS, MOMENTUM, AND ENERGY

working with two length scales in the y-direction, and it is the relation between them that we seek. We need to express the velocities in terms of $\hat{y}_T$. Now, we know that, to a rough approximation, $\hat{u} \sim y/\delta = \hat{y}_T(\delta_T/\delta)$. Since $\partial \hat{u}/\partial \hat{x} = -\partial \hat{v}/\partial \hat{y}$, and $\partial \hat{u}/\partial \hat{x} \sim \hat{u} \sim \hat{y}$, it then follows that $\hat{v} \sim -\hat{y}^2 = -\hat{y}_T^2 (\delta_T/\delta)^2$. Thus Eq. (6.36) can be written

$$\frac{\rho c_p U \delta_T^3}{kL\delta} \left(\hat{y}_T \frac{\partial \hat{T}}{\partial \hat{x}} - \hat{y}_T^2 \frac{\partial \hat{T}}{\partial \hat{y}_T} \right) \sim \frac{\partial^2 \hat{T}}{\partial \hat{y}_T^2} \tag{6.37}$$

The conduction and convection terms are then comparable if $\rho c_p U \delta_T^3/kL\delta$ is of order unity. This fixes δ_T; with Eq. (6.26) and the definition of the *Prandtl number*,

$$\Pr = c_p \eta / k \tag{6.38}$$

we then obtain

$$\delta_T = \delta \Pr^{-1/3} = L \operatorname{Re}_L^{1/2} \Pr^{-1/3} \tag{6.39}$$

The heat transfer coefficient, h, is defined as the heat flux from the plate divided by the temperature difference (cf. Eq. 6.8). Since v is zero right at the surface, the heat transfer into the fluid is by conduction, and we have

$$h = \frac{k(\partial T/\partial y)}{T_\infty - T_w} = \frac{k}{\delta_T} \frac{\partial \hat{T}}{\partial \hat{y}_T} \sim \frac{k}{\delta_T} \tag{6.40}$$

The *Nussult number* is defined

$$\operatorname{Nu}_x = \frac{hx}{k} \tag{6.41}$$

It then follows from Eqs. (6.39) and (6.40) that

$$\operatorname{Nu}_x \sim \operatorname{Re}_x^{1/2} \Pr^{1/3} \tag{6.42}$$

The result obtained by solving the partial differential equations is

$$\operatorname{Nu}_x = 0.332 \operatorname{Re}_x^{1/2} \Pr^{1/3} \tag{6.43}$$

This result is limited to $\Pr > \frac{1}{2}$, which excludes liquid metals. A different scaling argument can be used to obtain the appropriate result for small Prandtl numbers.

6.6.4 Some Conclusions

This problem can be carried much further. The fact that δ and δ_T vary as $x^{1/2}$, for example, and that $\hat{u}$, $\hat{v}$, and $\hat{T}$ depend on $y/\delta \sim y/x^{1/2}$, suggests immediately that the independent variable should be changed. The result is a set of *ordinary* differential equations in $y/x^{1/2}$. The related problem of mass transfer from a surface to a fluid is handled in identical fashion. We shall go no further, however, because the important points with regard to modeling have been made.

First, we see how nondimensionalization and the choice of the proper scaling variables can lead directly to important conclusions that might be adequate in themselves. The information contained in Eqs. (6.32) and (6.42) will often be all that is needed in the engineering analysis of a convective heat transfer problem, for example, and the substantial additional labor to solve the nonlinear partial differential equations may be unnecessary. This approach is too little used.

Second, we see the way in which scaling is used to simplify a model and reduce the complexity of the formulation. This may lead to a problem that is more tractable either analytically or computationally. The boundary layer equations (6.30) are much more tractable than the Navier-Stokes equations, for example. This process is the most common use of scaling, and it plays a central role in all analysis.

6.7 A LOGICAL PROBLEM

The procedure outlined above for model simplification is nearly always used (though rarely in so formal a manner), but its application does contain a possible problem. Let us suppose that our starting equations are, in some vague sense, an *exact* representation of the physical system. The logical process that we employ for problem solving is as follows:

1. Scale the variables so that the resulting dimensionless equations contain only variables of order unity.
2. Delete terms in the *exact* equations because variables of order unity are multiplied by a small parameter. (This is the process leading to the boundary layer equations, for example.)
3. Solve the resulting *approximate* equations. For the sake of the discussion, assume that this solution is an exact solution to the approximate equations.
4. Use the solution to estimate the magnitude of the neglected terms to ensure that they are truly negligible relative to terms that were retained.

6.7 A LOGICAL PROBLEM

The logic is shown in Fig. 6.3. We want an approximate solution to the exact equations. The logical process contains no mechanism for establishing that this is indeed what we have found. We assume that an exact solution to an approximate equation is an approximate solution to an exact equation. This is usually true, but need not be so.

Segal has developed the following illustration of the logical difficulty here. Consider the equations

$$0.01u + v = 0.1 \tag{6.44a}$$

$$u + 101v = 11 \tag{6.44b}$$

If we presume that the problem is properly scaled, and that u and v are of order unity, then the term $0.01u$ can be neglected relative to v. In that case we obtain $v = 0.1$ from Eq. (6.44a) and $u = 11 - (101 \times 0.1) = 0.9$ from Eq. (6.44b). The term that we neglected is then $0.01 \times 0.9 = 0.009$, which is indeed negligible when compared to $v = 0.1$. Thus, the approximation appears to be adequate. The exact solution, however, is $u = -90$, $v = 1$!

The problem has not, in fact, been properly scaled, although it is not evident that scaling alone would resolve the difficulty in a less transparent situation. Let U and V represent the scale factors, and define $\hat{u} = u/V$, $\hat{v} = v/V$. In order to see the problem structure more clearly, we will replace the coefficient 0.01 with a parameter ϵ. Thus, the scaled equations become

$$\frac{\epsilon U}{V} \hat{u} + \hat{v} = \frac{0.1}{V} \tag{6.45a}$$

$$\frac{U}{101V} \hat{u} + \hat{v} = \frac{11}{101V} \tag{6.45b}$$

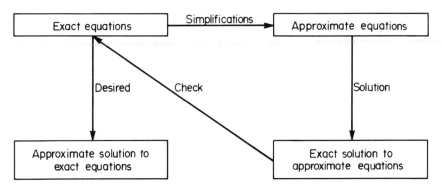

FIG. 6.3
Logic of approximate solutions.

From Eq. (6.45b), we see that all terms become comparable if $U = 101V$ and $V \sim 11/101$, so $U \sim 11$. The scaled equations are then

$$(101\epsilon)\,\hat{u} + \hat{v} = 0.91818\ldots \tag{6.46a}$$

$$\hat{u} + \hat{v} = 1.0 \tag{6.46b}$$

Clearly, in order to neglect the $\hat{u}$-term in the first equation, we require $101\,\epsilon \ll 1$; this inequality is not satisfied by $\epsilon = 0.01$, as in Eq. (6.44), so neglecting that term was improper. We see further that for $101\epsilon = 1$ there is no solution. Thus, we anticipate considerable sensitivity in the neighborhood of $\epsilon = 0.01$.

6.8 SENSITIVITY

The sensitivity of the solution of a set of equations to a parameter can sometimes be estimated. Sensitivity analysis is particularly useful for nonlinear systems, because the sensitivity equations are themselves linear. We will illustrate the procedure with just one special case of a pair of algebraic equations, and we will apply the result to the linear system in the preceding section. The same approach works for differential and integral equations.

Consider the pair of equations

$$f(u,v,\epsilon) = 0, \qquad g(u,v,\epsilon) = 0 \tag{6.47}$$

For the example in Eqs. (6.45) we have

$$f = \epsilon u + v - 0.1 \tag{6.48a}$$

$$g = u + 101v - 11 \tag{6.48b}$$

The solutions, u and v, can be considered to be functions of ϵ, $u(\epsilon)$ and $v(\epsilon)$. The sensitivity to the parameter is usually defined as $d\,\ell n\,u/d\epsilon$, $d\,\ell n\,v/d\epsilon$; this choice is made because the derivative of the logarithm reflects the fractional change in the solution for a unit change in the parameter.

Since Eqs. (6.47) are valid for all ϵ, we can differentiate with respect to ϵ to obtain

$$\frac{\partial f}{\partial u}\frac{du}{d\epsilon} + \frac{\partial f}{\partial v}\frac{dv}{d\epsilon} = -\frac{\partial f}{\partial \epsilon} \tag{6.49a}$$

$$\frac{\partial g}{\partial u}\frac{du}{d\epsilon} + \frac{\partial g}{\partial v}\frac{dv}{d\epsilon} = -\frac{\partial g}{\partial \epsilon} \tag{6.49b}$$

These are linear equations for $du/d\epsilon$ and $dv/d\epsilon$, so they are easily solved. The derivatives of f and g are evaluated at the solution.

Let us illustrate with Eqs. (6.48): $\partial f/\partial u = \epsilon$, $\partial f/\partial \epsilon = u$, etc. Thus, Eqs. (6.49) become

$$\epsilon \frac{du}{d\epsilon} + \frac{dv}{d\epsilon} = -u \tag{6.50a}$$

$$\frac{du}{d\epsilon} + 101 \frac{dv}{d\epsilon} = 0 \tag{6.50b}$$

The solution is

$$\frac{1}{u}\frac{du}{d\epsilon} = \frac{d\ln u}{d\epsilon} = \frac{101}{1 - 101\epsilon} \tag{6.51a}$$

$$\frac{1}{v}\frac{dv}{d\epsilon} = \frac{d\ln v}{d\epsilon} = -\frac{u}{v}\frac{1}{1 - 101\epsilon} \tag{6.51b}$$

We see that the sensitivity is extremely large for 101ϵ close to unity. We expect this, since the solution no longer exists at $101\epsilon = 1$.

6.9 CONCLUDING REMARKS

Scaling is one of the most powerful tools available to the analyst. It plays a central role in modeling at every stage. We have just touched here on the uses. Most modern asymptotic methods for solving complex equations are dependent on proper scaling, since perturbation methods are generally employed.

Dimensional analysis and scaling seem to be much less appreciated now than a few decade ago. This is probably a negative consequence of the increased use of digital computers and, recently, calculators and personal computers. Numerical solutions can be obtained to rather complex equations, obviating the need for the order-of-magnitude estimation and approximation inherent in older analytical methods and use of the slide rule. The analog computer required proper scaling or a program would not work; analog computers only allow variables to be represented by voltages between $\pm$ 10 V. This type of precomputational thinking and consequent simplification does not always seem necessary (although it is!) when working with a machine that computes over ninety-nine orders of magnitude.

BIBLIOGRAPHICAL NOTES

For a nice discussion of scaling, see

C. C. Lin and L. A. Segal, *Mathematics Applied to Deterministic Problems in the Natural Sciences* (New York, NY: Macmillan, 1974).

Aris develops the subject of nondimensionalization in a somewhat different manner from that developed here, and includes a large number of references:

R. Aris, *Mathematical Modelling Techniques* (London: Pitman, 1979).

The equations in Table 6.1, with the exception of the energy equation, are derived in most books on fluid mechanics. All of the equations are derived rigorously in

R. B. Bird, W. E. Stewart, and E. N. Lightfoot, *Transport Phenomena* (New York, NY: John Wiley, 1960).

The scaling arguments used for the Navier-Stokes equations follow the development in

M. M. Denn, *Process Fluid Mechanics* (Englewood Cliffs, NJ: Prentice-Hall, 1980).

The corresponding scaling arguments for the heat transfer do not seem to have been published, although the result is part of the "folklore" of heat and mass transfer. Oistrach uses a similar methodology. See, for example, S. Oistrach, *Ann. Rev. Fluid Mech.* **14** (1982) 313.

An approach similar to the one used here, with a number of interesting applications and extensions to more complicated situations, is described in

E. Ruckenstein, in *Handbook for Heat and Mass Transfer Operations* (Houston, TX.: Gulf Publishing Co., 1985), p. 83.

Solutions to the boundary layer equations may be found in Bird *et al.*, Denn, and most books on heat transfer and fluid mechanics.

7

CASE STUDY: COAL GASIFICATION

7.1 INTRODUCTION

Coal gasification reactors are devices that contact coal with reactive gases, usually oxygen and steam, in order to produce gaseous products. The major chemicals produced are hydrogen, carbon monoxide, and methane. The desired products might be fuel gas, SNG, or syngas; each of these products requires a different gas composition, and hence a different reactor design or feedstock.

Fuel gas is a clean fuel that can directly substitute for natural gas in existing boilers and other industrial applications, and is suitable for use in combined-cycle power plants. The major criterion here for evaluating performance is the lower heating value of the gas, which is approximately equal to the number of moles of $CO + 0.85\ H_2 + 2.85\ CH_4$ produced per mole of coal. A high CO content is therefore preferable. The major disadvantage of this gas, which typically has a low methane content, is that it is expensive to pump over long distances.

SNG (substitute natural gas, sometimes "synthetic natural gas") is methane produced from a coal gasifier and further downstream processing, typically from a Fischer-Tropsch process. The total yield of methane per mole of coal is proportional to $CO + H_2 + 4CH_4$ in the product gas. A gasifier that is part of an SNG process therefore will ideally maximize production of methane in order to reduce gas cleaning and methanation costs;

the ratio H_2/CO in the gasifier product should be greater than unity for the methanation process.

Syngas is the name used for mixtures of CO and H_2 used in the manufacture of a variety of chemicals. A particular H_2/CO ratio is required for each application. In all cases, the methane formed in the gasifier is a by-product that must be separated. (This does not imply that methane production should necessarily be minimized in a syngas plant, since gasifiers with high thermal efficiency always produce significant amounts of methane, and eliminating methane production imposes severe design penalties that must be weighed against the cost of methane separation.)

A number of gasifier configurations are currently in use throughout the world, and others are under development. The *moving bed gasifier* was described briefly in Section 2.4. The prototype of this class of gasifiers is the Lurgi dry-ash pressurized reactor, shown schematically in Fig. 2.3, which has been in commercial operation since 1936 and which is similar in concept to nonpressurized town gas producers that have long been in use. Many of the operating problems for this type of gasifier are associated with mechanical devices and bulk measurement of solids, and are not appropriate for discussion here; two processing problems that have been the focus of considerable modeling effort, and which we will consider in this chapter, are the control of the maximum temperature in the reactor and the estimation of long- and short-term transient response characteristics.

7.2 OVERALL CONSTRAINTS

Good engineering always entails a preliminary analysis to seek broad bounds on behavior with minimal effort. Scaling arguments of the type utilized in the early sections of Chapter 6 often serve this purpose. So, too, does application of the conservation laws to a large control volume, typically comprising the entire piece of process equipment. Such an overall, or *stoichiometric,* analysis requires little or no use of constitutive relationships.

In analyzing the moving bed gasifier, we will first consider only the gasification of char, which can be taken in the first approximation as consisting only of carbon. As will become evident, any effect of the presence of volatiles can be added afterwards in this overall analysis. We shall also assume, in the first approximation, that the reactions of char to form methane are much slower in the absence of a suitable catalyst than char combustion or the reaction of char with steam or CO_2, and that we may therefore tentatively remove methane from the list of possible reaction products. All other reactions to form CO, CO_2, and hydrogen from carbon, water, and oxygen can be constructed from linear combinations of the following three

7.2 OVERALL CONSTRAINTS

linearly independent chemical reactions:*

$$C + \tfrac{1}{2}O_2 \rightleftarrows CO \qquad \Delta H_1 = -26.4 \qquad (7.1)$$

$$C + O_2 \rightleftarrows CO_2 \qquad \Delta H_2 = -94.2 \qquad (7.2)$$

$$C + H_2O \rightleftarrows H_2 + CO \qquad \Delta H_3 = +32.2 \qquad (7.3)$$

ΔH shown for each reaction is the enthalpy change ("heat") of reaction, which is negative for an exothermic reaction.

The essential inputs to a coal gasification reactor are carbon, oxygen, and water. The coincidence of having three compounds allows a convenient graphical representation of reactor operation through the use of triangular coordinates. Consider Fig. 7.1. The coordinates of each point sum to unity; point A, for example is $0.58C$, $0.16O_2$, and $0.26H_2O$. Thus, each point can be considered to represent a potential set of feed conditions to the gasifier, with the actual operating conditions obtained by multiplying by flow rate. Feed compositions corresponding to reactions (7.1) through (7.3) are shown on the axes. Incomplete carbon conversion must occur for starting compositions that lie above the line connecting $C + \tfrac{1}{2}O_2$ and $C + H_2O$, since there is insufficient reactant to utilize all the carbon, while incomplete oxygen conversion will occur for compositions below the line connecting $C + O_2$ to the H_2O vertex. Steam generation is costly; it can be shown by linear combination of the basis reactions [twice (3) plus (2) minus twice (1)] that steam must appear in the effluent for starting compositions that lie below the line connecting $C + O_2$ with the reaction

$$C + 2H_2O \rightleftarrows CO_2 + 2H_2, \qquad \Delta H = +23.0$$

(Some steam will appear in the effluent for starting formulations above this line, since these bounds are based on complete conversion and neglect any possible chemical equilibria.) The shaded area thus bounds the most desirable region for gasifier operation.

The operating range can be further restricted by the condition of autothermal operation (overall thermal neutrality): the exothermic reactions must produce just sufficient reaction enthalpy to drive the endothermic reactions. We suppose that the inlet and outlet streams for char gasification are at the same temperature. Let X_1, X_2, and $1 - X_1 - X_2$ represent the fractional

---

*The stoichiometry of chemical reactions can be studied using linear algebra. The number of independent reactions that must be considered equals the rank of a reaction matrix; the rank is usually the difference between the number of chemical species and the number of elements from which those molecules are constructed.

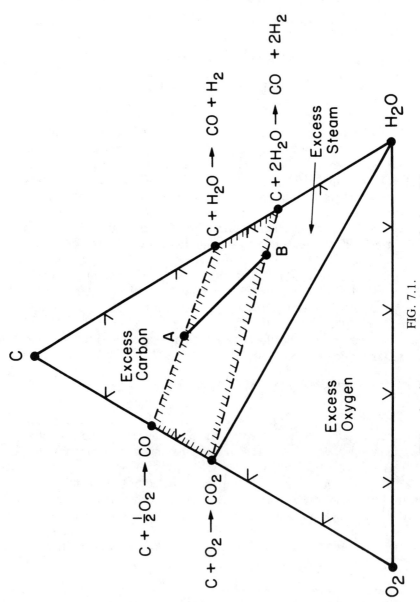

FIG. 7.1. *Triangular diagram representing reactor feed conditions.*

conversion of carbon through each of the basis reactions (1), (2), and (3), respectively. Thermal neutrality is then represented by

$$(-26.4)X_1 + (-94.2)X_2 + (+32.2)(1 - X_1 - X_2) = 0$$

The overall reaction for thermal neutrality is the one-parameter family

$$C + (0.255 + 0.036X_1)O_2 + (0.745 - 0.536X_1)H_2$$
$$\rightarrow (0.745 + 0.464X_1)CO + (0.255 - 0.464X_1)CO_2$$
$$+ (0.795 - 0.536X_1)H_2$$

This reaction family lies along line AB in Fig. 7.1. Point A corresponds to $X_1 = 0.55$; X_1 cannot exceed 0.55, or else CO_2 will be required as a feed. Point B corresponds to $X_1 = -1.61$ and represents the minimum value for which it is possible in principle to produce a product gas without steam. Values of X_1 below -1.61 would require CO as a feed in this simplified analysis. The H_2/CO ratio in the product gas varies from 0.45 at point A to infinity (no CO) at point B. Conversely, point A has no CO_2 in the product, while the H_2/CO_2 ratio at point B is 2. While the nominal thermal efficiency of all points along line AB is 100%, based on lower heating value, this does not take into account the energy required to produce the steam and oxygen. The production of one mole of oxygen at 400 psi and 100° F requires about four times the energy to produce one mole of steam at the same conditions; on this basis, point A is more efficient than point B.

An equivalent analysis can be carried out by assuming that methanation will take place, and that the reaction products are only CO, CO_2, and CH_4. This limit is unrealistic in the absence of the appropriate catalysts, but helps in defining the performance that might be obtained. Combination of the two bounding analyses leads to the operating diagram shown in Fig. 7.2. The quadrilateral $ABEF$ bounds the autothermal region and lies within the region of possible complete utilization of carbon and steam. If methanation could be carried out within the gasifier, then the most efficient operation in terms of costs of steam and oxygen production would be along line EF. Kinetic constraints preclude this option, however, and it is therefore clear that the desirable region for operation of a gasification reactor is close to point A. (The assumption that all reactions go to completion is not as restrictive as it might appear. The oxygen and steam utilized to convert carbon must satisfy the constraints outlined here, which thus define upper bounds on efficiency. Equilibrium, kinetic, and transport limitations provide information regarding excess steam. Oxygen utilization will always be essentially complete.)

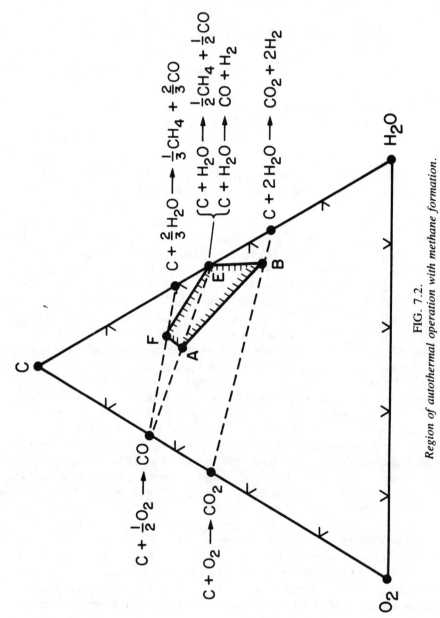

FIG. 7.2.
Region of autothermal operation with methane formation.

7.3 KINETICS-FREE MODELING

The "kinetics-free" approach to be described here is an application of scaling principles that is particularly useful in the modeling and analysis of chemical reactors. The scaling is crude, and a formalism like that developed in the preceding chapter need not be introduced. Rather, use is made of the fact that the rates of some chemical reactions are such that little or no conversion will occur over a time scale of the order of the residence time in the reactor, while the reactor is likely to be designed with a residence time adequate to allow other, faster reactions to proceed to completion or to equilibrium, as appropriate. The term *kinetics-free* is a misnomer in this sense, since relative values of reaction and residence times are explicitly accounted for, but these ratios are taken in this approximation to be either very small or very large.

When employing this approach it is useful to replace the second combustion reaction, Eq. (7.2), with the water-gas shift reaction as a linearly independent reaction,

$$CO + H_2O \rightleftarrows CO_2 + H_2, \quad \Delta H = -9.2 \qquad (7.4)$$

[Equation (7.4) is obtained by adding Eqs. (7.2) and (7.3), then subtracting (7.1).] This reaction is quite fast, and can normally be assumed to go to equilibrium at the temperature in a gasifier. Direct hydrogenation to form methane is usually slow, and will be neglected initially in this elementary analysis. Oxygen reactivity is assumed to be infinitely fast, and oxygen attack on the fixed carbon is assumed to be irreversible and completed before any other reaction begins. Steam gasification, reaction (7.3), is taken to be sufficiently fast to reach equilibrium or completion at the top of the reaction zone. If the coal feed rate is too large to utilize all of the steam, then reactions (7.3) and (7.4) will be in equilibrium at the top of the reaction zone and there will be unreacted carbon in the ash at the bottom of the reactor. Reduction in the carbon feed rate relative to steam will lead to reduced amounts of carbon in the ash while maintaining the equilibria at the top. The carbon in the ash will reach zero at a critical carbon rate. At carbon feed rates below this critical value there will be an excess of steam and the carbon-steam reaction can no longer be at equilibrium at the top of the gasification zone, although the equilibrium of reaction (7.4) will be maintained. (This behavior is contrary to the intuition of many people and is a consequence of the countercurrent operation.)

Carbon conversion is specified, all feed oxygen is assumed to be consumed, and solid and gas temperatures are assumed to be equal. The overall mass and energy balances, together with the equations describing the chemical equilibria,* are then sufficient to define the composition and temperature of the gas above the gasification zone, but prior to devolatization and drying. Because the two oxidation reactions have significantly different reaction enthalpies, the distribution of reaction products between CO and CO_2 must be specified if an estimate of the maximum temperature is desired. The computation then proceeds as follows, starting from the reactor bottom with a specified exit carbon composition:

1. Combustion utilizes all feed oxygen. The composition above the combustion zone is fixed by the distribution of oxidation products between CO and CO_2, and the temperature is computed from the adiabatic temperature rise. (Heat losses may be included if the region near the reactor wall is to be accounted for, but an adiabatic analysis is adequate for a reactor like a Lurgi gasifier with a 3 to 4 m diameter.) The fixed-carbon to ash ratio in the coal must be specified, since the heat capacity of the ash enters into the energy balance. The calculations shown here use a ratio of 5:1, as typical of Illinois No. 6 coal.
2. Water-gas shift equilibrium is established adiabatically above the combustion zone. The temperature computed here is a very rough estimate of the maximum temperature in the reactor. (Steps 1 and 2 can, of course, be combined.)
3. The remaining carbon is converted by reaction with steam. The composition above the gasification zone is fixed by the carbon reaction with steam, and the temperature is computed from the adiabatic temperature decrease. (Heat losses may be included.) Only the steam reaction is needed, since water-gas shift equilibrium will be established and the CO_2 reaction is not independent.
4. Water-gas shift equilibrium is established adiabatically above the gasification zone. This is the composition and temperature at which product gas leaves the reaction zone of the gasifier. (Steps 3 and 4 can also be combined.)

The final gas temperature and composition are fixed by the overall mass and energy balances and the water-gas shift equilibrium, together with the requirements of complete oxygen and specified carbon utilization. Thus,

---

*The equilibrium relationships for Eqs. (7.3) and (7.4) are, respectively,

$$K_3(T) = p \frac{y_{H_2} y_{CO}}{y_{H_2O}}, \qquad K_4(T) = \frac{y_{CO_2} y_{H_2}}{y_{CO} y_{H_2O}}$$

where y_i are mole fractions in the gas phase and p is the total pressure. The temperature-dependent equilibrium "constants" are tabulated in the literature.

7.3 KINETICS-FREE MODELING

the effluent properties are independent of the initial oxidation CO/CO_2 split and particle size. Indeed, if an estimate of the maximum temperature is not required, the complete calculation can be carried out in one step.

The calculations shown in Fig. 7.3 are for a gas-feed temperature of 700° F and an operating pressure of 25 atm. The solid lines are lines of constant fraction of unreacted carbon in the ash. The line of zero unreacted carbon represents the case in which feed rates of carbon, steam, and oxygen are such that the carbon-steam reaction reaches equilibrium at the top of the reaction zone without unreacted carbon at the bottom; this line and all points below are characterized by complete utilization of carbon, first by oxygen and then by steam.

There is a noticeable change in slope of the lines of constant unreacted carbon at a steam-to-oxygen molar ratio of about 1.2. To the left of the change in slope the equilibrium of the carbon-steam reaction is far towards the formation of CO and H_2, and this reaction as well as oxidation can be taken to be essentially irreversible. The line of complete carbon conversion in this region is thus very close to the line connecting reactions (7.1) and (7.3) in Figs. 7.1 and 7.2. The region to the right of the break represents an excess of steam. The excess steam lowers the temperature, and sufficient excess steam brings the peak temperature below the ash melting point and allows operation of a dry ash reactor. The lines of constant carbon in the ash in this region approximate lines of constant oxygen-to-carbon molar feed ratios.

The dashed lines in Fig. 7.3 are lines of constant increase in gas sensible heat in Kcal/g-mole of fixed carbon. Autothermal operation for char gasification lies to the right of the line of complete carbon conversion. A coal containing 10% moisture will require about 5 Kcal/g-mole of carbon for heating and drying; if we presume that the gas exit temperature must be at least 700° F to prevent condensation of tars, then operation in the region to the right of the 5 Kcal/mole line is excluded because of the need to add heat to the gasifier. Operating points to the left of the 5 Kcal line will represent an appreciable loss of energy unless the sensible heat can be conveniently recovered downstream, which will be difficult in a gas stream containing condensibles that will foul a heat exchanger. The intersection of the 5 Kcal/mole sensible heat line with the complete carbon conversion line thus represents the most efficient operating point. (This is slightly to the left of point A in Fig. 7.1, which was determined on the basis of a dry, hot char feed.)

There is no conceptual or computational difficulty in repeating these calculations with a reaction to form methane included in the basis. The line of complete carbon conversion with methanation equilibrium is shown in Fig. 7.4, together with the corresponding line without methanation. The two lines converge at steam-to-oxygen ratios of order unity, which is the region

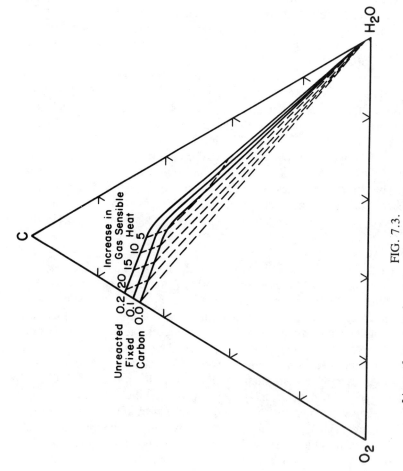

FIG. 7.3. Lines of constant fraction of unreacted fixed carbon (solid) and increase in gas-sensible heat (dashed, Kcal/mole fixed carbon) for moving bed gasifier (after Yoon et al., EPRI AF-590).

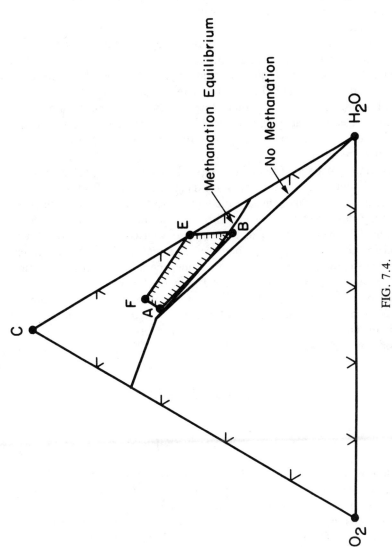

FIG. 7.4.
Lines of complete carbon conversion with and without methanation for moving bed gasifier (after Yoon et al., EPRI AF-590).

136 *7 CASE STUDY: COAL GASIFICATION*

of operation of the slagging moving bed. The high temperatures here drive the equilibrium far to the left and suppress methane formation. Thus, for slagging operation the methanation reactions need not be considered. At the higher steam-to-oxygen ratios and lower temperatures characteristic of the dry ash Lurgi moving bed reactor, however, methanation is thermodynamically favorable.

The trapezium of autothermal operation in the reaction zone (*ABEF*) is also included in Fig. 7.4. The neighborhood of line *EF*, which is carbon conversion to methane, CO, and CO_2 and contains the highest efficiencies, is quite far removed from the line of complete carbon conversion with methane equilibrium. This region around line *EF* therefore appears not to be obtainable. Selective catalysts that provide a kinetic pathway approaching global equilibrium by a trajectory passing through the neighborhood of line *EF* would be required.

A kinetics-free analysis of this type clearly gives a great deal of information about the reactor operation in exchange for little effort, and should always be an early component of a modeling effort. (Such was not the case in the study described here, where this analysis in fact followed a more detailed model development! This good advice is based on hindsight.) Some essential information is missing, however. The calculation requires an estimate of the carbon conversion, whereas the operating conditions leading to good carbon conversion might not be known *a priori*, and can only be predicted by a reactor model that utilizes rate information about chemical and physical processes. Such a model is also required for calculation of the temperature profile, which, we have noted, is of primary concern from an operational viewpoint.

7.4 DETAILED MODELING

7.4.1 Broad Principles

A detailed model of the reactor, based on a differential control volume, is required in order to compute the spatial temperature distribution (and hence the location and magnitude of the temperature maximum), as well as the carbon conversion. (The latter, it will be recalled, is an input to the kinetics-free calculation.) The overall approach is not different in concept from the examples in Chapters 4 and 5, but there are important differences in detail that make the modeling task for the moving bed gasifier a difficult one. At least eleven detailed models of varying degrees of complexity, and with different physical assumptions, have been described in the published literature.

First, we must decide whether to include the radial variations that are

7.4 DETAILED MODELING

induced by the wall cooling jacket. The scaling estimate in Section 6.5, based on the Einstein diffusion time, suggests that radial effects will be limited to a region within about 7 cm from the wall. This is not a particularly large portion of the cross section of the 4 m to 5 m diameter in a commercial reactor, and radial effects are neglected to the first approximation. (We shall return to this issue later in the chapter, since radial effects should be important for a reactor with a smaller diameter, or when the throughput is substantially reduced from the figures on which the scaling in Section 6.5 was based.)

The most fundamental modeling decision has to do with the treatment of the temperatures of the two phases. For small particles the rate of heat transfer within the solid will be quite rapid (the Einstein diffusion time!) and the two temperatures will be essentially the same. Heat transfer resistance between the gas and solid could slow the heating rate for 1-cm diameter particles, however (a small Biot number, Section 6.3), and there could be a temperature difference between the solid and gas. If the temperatures are equal, then it is possible to consider the gas and solid as a single "pseudo-phase," with average properties, as was done for the FCC in Section 5.9. If the temperature difference cannot be neglected, then differential control volumes must be constructed in each of the individual phases, greatly increasing the complexity of the problem, particularly the numerical computation. Both approaches have been used; temperature differences do not seem to be large over most of the length of the reactor, and the modeling discussed here is based on the assumption of equal solid and gas temperatures.

The other important modeling decision is the nature of the description of the solid-gas reactions and the degree of detail used to describe the physicochemical phenomena in the neighborhood of a coal or char particle. Considerable attention has been given in the literature to the problem of combined mass and heat transfer and chemical reaction in the thin gas layer about a reacting particle, with the (sometimes) goal of developing a rate constitutive equation describing the appearance and disappearance of chemical species that is appropriate for use in a continuum model of the reactor (i.e., one assumes, as in Section 5.9, that the reaction takes place at every spatial point, and not only at the discrete locations of solid particles). The simplest form is used in the modeling described here; there are three resistances in series: mass transfer of reactants through an external gas film to the surface of the particle, diffusion of reactants through an ash layer to the surface of the unreacted carbon, and chemical reaction with the carbon.*

---

*The term $r_{i,\text{reaction}}$ is itself confounded by the fact that reactants may penetrate the carbon and react in the interior as well as at the surface; this is accounted for by use of an "effectiveness factor," which adjusts the intrinsic chemical rate to account for the fact that all of the carbon is not accessible to the reactants.

138 *7 CASE STUDY: COAL GASIFICATION*

FIG. 7.5.
Coal particle removed from ash lock in Lurgi gasifier, demonstrating shell-progressive structure. (Photo by G. M. Whitney)

The overall rate of the ith gas-solid reaction, r_i, is then written in terms of each of these steps as

$$\frac{1}{r_i} = \frac{1}{r_{i,\text{ external transfer}}} + \frac{1}{r_{i,\text{ ash diffusion}}} + \frac{1}{r_{i,\text{ reaction}}}$$

(This reciprocal additive form is clearly correct, in that the slowest rate step determines the overall rate.) Figure 7.5 is a photo of a piece of coal removed from the ash lock of a Lurgi gasifier in Hyderabad, India, and shows the appropriateness in this case of the "shell progressive" model of the particle conversion. Another limiting case would be the "ash segregation" single-particle model, in which the ash layer is removed immediately upon reaction and is not present on the coal particle; the ash diffusion contribution does not then appear in Eq. (7.5).

7.4.2 One-Dimensional Homogeneous Model

The reactor model described here is often known in the literature as the UD/EPRI (for University of Delaware/Electric Power Research Institute) model;

7.4 DETAILED MODELING

the similar LBL/DOE (for Lawrence Berkeley Laboratory/Department of Energy) model relaxes some assumptions and is available in a far more efficient computer code. While the end products (the computer codes) differ significantly, there are few essential differences in modeling and they are not differentiated here. Reactions (7.1) through (7.4) are assumed to occur for char, together with the following two reactions:

$$C + CO_2 \rightleftarrows 2CO, \quad \Delta H_5 = 41.4 \tag{7.5}$$

$$C + 2H_2 \rightleftarrows CH_4, \quad \Delta H_6 = -20.2 \tag{7.6}$$

The reaction $H_2 + \frac{1}{2}O_2 \rightleftarrows H_2O$ is assumed *not* to occur; this assumption is controversial, but can be defended. The fixed carbon in the char is assumed to contain only carbon.

Let z denote distance measured from the bottom of the reactor, and let $G(z)$ denote the total number of moles of gas per unit area of reactor. A typical component mass balance at steady state, say on water, then proceeds as follows:

$$\text{time rate of change} = 0 = \underbrace{Gy_{H_2O}|_z}_{(A)} \underbrace{-Gy_{H_2O}|_{z+\Delta z}}_{(B)} \underbrace{-r_3(1-\epsilon)\Delta z}_{(C)} \underbrace{-r_4 \epsilon \, \Delta z}_{(D)}$$

y_{H_2O} is the mole fraction of steam in the gas phase. The terms (A) and (B) represent the respective molar flow rates (per unit area) of steam into and out of a control volume of height Δz. (C) represents the loss of steam through reaction (7.3); ϵ is the volume fraction that does not contain solids (the *void fraction*), and the term $(1 - \epsilon)$ is included to account for the fact that r_3 will be defined per unit volume of solid. Similarly, (D) represents the loss rate of steam through the gas-phase water gas shift reaction, and the void fraction is included to account for the fact that the reaction rate is defined on the basis of a unit volume of gas. (The UD/EPRI model assumes that the water-gas shift reaction goes to equilibrium, leading to a slight but unimportant change in the formulation.) Axial diffusion of mass is assumed to be unimportant; this assumption is easily supported using scaling arguments. Dividing by Δz, and taking the limit as $\Delta z \to 0$, we then obtain the differential equation

$$\frac{dGy_{H_2O}}{dz} = -(1 - \epsilon)\, r_3 - \epsilon r_4 \tag{7.7}$$

Similar equations follow for the other gaseous species.* The overall balance

---

*Other forms of those equations can be obtained by variable manipulations, and the coding of both the UD/EPRI and LBL/DOE programs is based on derived forms of the mass balance equations.

for G has the form

$$\frac{dG}{dz} = (1 - \epsilon)\left[\tfrac{1}{2} r_1 + r_3 + r_5 - r_6\right] \tag{7.8}$$

since reaction (7.1) results in a net gain of one-half mole of gas, (7.2) and (7.4) cause no change in the number of moles, while (7.5) causes a net increase of one mole and (7.6) a net decrease of one. The total molar gas flux (steam plus oxygen feed) and the mole fractions of each gaseous species are given at $z = 0$. The carbon mass balance is most conveniently written in terms of the molar feed rate of fixed carbon per unit area of reactor, FC, and the fraction w of original fixed carbon that remains unconverted:

$$FC \frac{dw}{dz} = (1 - \epsilon)(r_1 + r_2 + r_3 + r_5 + r_6) \tag{7.9}$$

w is, of course, unity at the *top* of the gasifier.

The derivation of the steady-state energy balance requires accounting for the separate enthalpy flows of solid and gas, and then utilizing the chain rule as in Section 5.3 to account for the composition dependence of the enthalpy. Partial molar enthalpies are grouped to form the enthalpies of reaction. Axial conduction of heat is neglected; this assumption is also easily supported using scaling arguments at full throughput, but fails when convective and conductive teams become comparable at about 10% of full design load. The energy equation has the final form

$$\left\{FC\left[(1-w) + \frac{f_A}{f_{FC}}\right] c_{ps} - G \sum y_i c_{pi}\right\} \frac{dT}{dz} = (1-\epsilon) \sum_{i \neq 4} r_i \Delta H_i + \epsilon r_4 \Delta H_4 \tag{7.10}$$

c_{ps} is the (equal) heat capacity of the solid carbon and ash, while c_{pi} refers to the heat capacity of the ith gaseous species. f_A and f_{FC} are fractions of ash and fixed carbon, respectively, in the proximate analysis. The negative sign between the two terms multiplying dT/dz is a consequence of the countercurrent flow; the equation would become singular if these convective heat capacity terms approached one another, and conduction would then have to be taken into account. The temperature is given at $z = 0$. The *boundary layer–adiabatic core* version of the model visualizes a second reactor, concentric to the adiabatic reactor described by Eqs. (7.3) through (7.10), which is nonadiabatic and exchanges heat with the jacketed wall but not with the adiabatic core. This annular reactor has a thickness equal to the Einstein diffusion length (7 cm), and is described by the same equation except for the addition to Eq. (7.10) of a heat transfer term of the form $ha(T - T_w)$.

7.4 DETAILED MODELING

h is a heat transfer coefficient, a is the heat transfer surface per unit volume of the heat transfer boundary layer region, and T_w is the wall temperature.

The thermodynamic quantities ($\{\Delta H_i\}$, c_{ps}, $\{c_{pi}\}$, equilibrium coefficients) are tabulated in standard references and are readily available. The several contributions to the constitutive equations for the reaction rates must be supplied. The rate of external mass transfer is expressed in terms of a standard correlation for the mass transfer coefficient between gas and particles in a packed bed. The rate of diffusion through the ash requires a diffusion coefficient; this is conventionally calculated as the product of the gas-phase diffusion coefficient with the square of the void fraction in the solid; the latter needs to be estimated. The intrinsic chemical reaction rates of gaseous species with solid carbon are taken to follow *mass action* kinetics, in which the rate is proportional to the partial pressure (mole fraction times total pressure) of each reacting species, with temperature-dependent coefficients. (Only the rate of the forward reaction is required; the reverse reaction rate is then fixed by the equilibrium coefficient.)

Some typical reaction rate data for high reactivity Wyoming and low reactivity Illinois No. 6 coals are shown in Fig. 7.6 as a function of temperature. The rate of the carbon–CO_2 reaction is about 0.6 that of carbon–H_2O. The first point to be made is that the scatter in the raw data is substantial, and different investigators differ on the correct value of the slope of the carbon-steam line; there is thus considerable uncertainty in the rate data. A single combustion line is shown, since the combustion rate is less sensitive to coal type than are the gasification rates. Furthermore, reliable data under dry ash gasifier conditions are available only for the *total* combustion rate, which is the sum of the rates of reactions (7.1) and (7.2); the *selectivity* (the CO/CO_2 ratio in the product gas) is not known. This selectivity is the single model input that is not based on independent literature data. Finally, the rate of the mass transfer steps in the overall reaction rate is shown for a 100-mm particle at 2000° F and 50% carbon conversion, which is typical of the combustion zone in a moving bed gasifier. Under these conditions the effective rates of the combustion and gasification reactions will be the same.

The modeling of the other important processes in the gasifier is quite crude. The drying time for a coal particle can be estimated to be of the order of one minute, which is a small fraction of the approximately one-hour solid residence time, and drying is therefore taken to be instantaneous at the top of the gasifier. Volatile evolution is also taken to be instantaneous at the top of the gasifier, with the assumption that the portion of the coal to be devolatilized is the same as the fraction of volatiles in the proximate analysis. Based on reported reactor data, it is assumed that 20% of the volatile matter appears as tar and oil plus phenol, while the remainder (of chemical composition $CHO_{0.05}$) is cracked to stoichiometric amounts of methane, car-

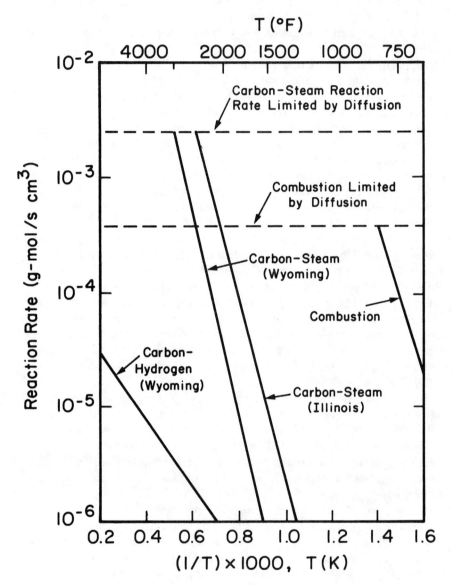

FIG. 7.6.
Rates of combustion and gasification reactions.

bon monoxide, and hydrogen. All of the sulfur in the coal is assumed to be converted to H_2S. The details of treating volatiles have little impact on the reactor operability, since volatile evolution takes place above the region where chemical reactions are occurring, though they are important in establishing the heating value of the product gas.

7.4.3 Model Results

Numerical solution of the model equations is straightforward, and can be carried out using conventional library programs for ordinary differential equations. (An integration routine for "stiff" differential equations helps, but is not essential). The boundary value problem is solved by assuming the amount of unreacted fixed carbon in the ash ($w(0)$) and then integrating the system equations to the top of the reaction zone (taken to be 98% of the reactor height), iterating on $w(0)$ until $w = 1$ at the top.

Figures 7.7 and 7.8 show computed temperature and composition profiles in the adiabatic core for conditions corresponding to runs on a Lurgi gasifier at Westfield, Scotland, using Illinois No. 6 coal. The computed effluent is tabulated in Table 7.1, along with the plant data. Effluent compositions are the only data available. The agreement is extremely good, given the uncertainty in the literature data and the fact that the only adjustable parameter in the model is the selectivity of the oxidation reactions. The maximum temperature (which cannot be measured) is also slightly below the ash softening temperature, as would be expected in an efficiently operating dry ash gasifier. The model could undoubtedly be "fine-tuned" by small changes in some parameters, but the errors in compositions are within the uncertainty of the purity of the oxygen feed to the reactor. Table 7.1 also shows results of a model by Cho and Joseph that is essentially equivalent in all respects in the reaction zone except that temperature differences are allowed between gas and solid, with separate energy balances for each phase. As we shall establish later, the differences between the results of the two models are most likely a consequence of different ways of handling the volatiles, which is the source of essentially all of the methane in both simulations. In fact, all moving bed models that have been compared with plant effluent data have shown generally good agreement.

7.4.4 Engineering Application

The model can be used to explore important processing considerations. Figure 7.9, for example, shows the sensitivity of the thermal efficiency for Illinois No. 6 coal to the oxygen/carbon feed ratio at a molar steam-to-

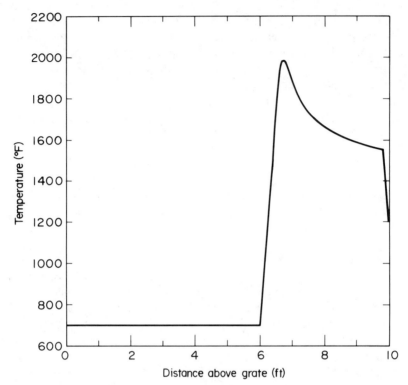

FIG. 7.7.
Computed temperature profile for conditions of Westfield tests (Table 7.1).

oxygen ratio of 9.6 (Westfield conditions). There is a sharp maximum at a feed ratio of slightly below 3; this corresponds to about 10% less oxygen than was used at Westfield. The dashed line in Fig. 7.10 is the locus of maxima in the thermal efficiency; it lies very close to the line in Fig. 7.3 computed with the kinetics-free approach for complete carbon conversion and no methanation. Computed lines of maximum temperature are also plotted in Fig. 7.10, completing the definition of the operability region. The dry ash reactor must operate to the right of the 1175° C line, which represents the softening temperature for the Illinois coal, but as close to that line as possible. Point L is the operating point for the Westfield tests in Table 7.1, and, as noted before, is consistent with the reactor model calculations. A slagging moving bed gasifier would have to operate to the left of the ash melting temperature, approximately 1500° C, and ideally as close to Point A as possible. It is clear that the feasible operating range for either type of moving bed gasifier is quite small.

Efficient operation of the gasifier requires that the carbon/oxygen

7.4 DETAILED MODELING

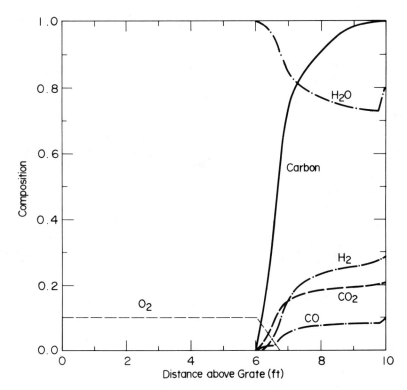

FIG. 7.8.
Computed composition profiles for conditions of Westfield tests (Table 7.1).

TABLE 7.1
Simulation of pressurized Lurgi gasification of Illinois No. 6 coal at Westfield; steam/oxygen molar ratio = 9.6, fixed carbon/oxygen molar ratio = 2.73.

	UD/EPRI Model	Cho and Joseph	Westfield Plant Data
CO_2	32.9	27.5	31.2
CO	13.8	19.5	17.3
H_2	42.4	44.4	39.1
CH_4	8.6	6.5	9.4
C_2H_6	0.7		0.7
H_2S	1.0	2.1	1.1
Inert	0.5		1.2

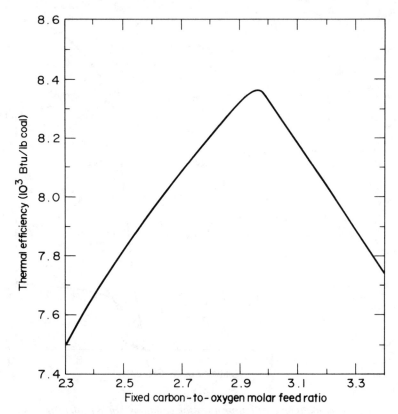

FIG. 7.9.
Thermal efficiency for gasification of Illinois No. 6 coal as a function of oxygen/carbon feed ratio (after Yoon et al., EPRI AF-590).

ratio be kept close to the optimum. As shown in Fig. 7.11, the position of the maximum temperature above the grate is a sensitive indicator, but it is impossible to make a direct measurement of the core temperature profile under operating conditions. The unreacted carbon in the ash is an effective indicator of insufficient oxygen, as shown in Fig. 7.12. It is interesting to note that most of the unreacted carbon comes from the cold region near the wall. The CO_2 content of the product gas is often used as a control variable. Model calculations for effluent CO_2 for both Illinois and Wyoming coals are shown in Fig. 7.13. This is clearly an effective control variable for the low reactivity Illinois coal, but not for the high reactivity Wyoming coal.

Other, similar calculations relating operating variables to performance can be (and have been) carried out, establishing a rather complete picture of reactor operability. The model is clearly inadequate in many re-

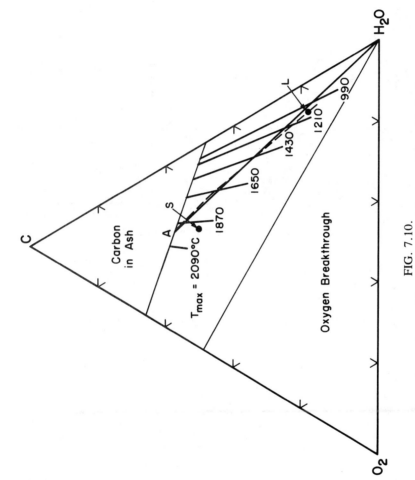

FIG. 7.10.
Operating region and locus of optimal feed conditions (dashed) computed for moving bed gasifier (after Denn and Wei, 1982).

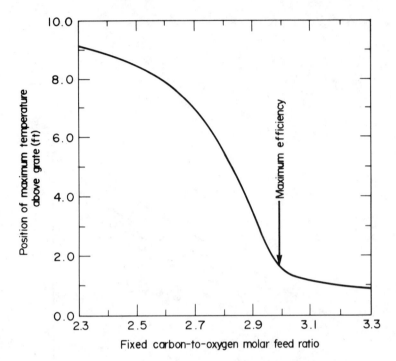

FIG. 7.11.
Position of maximum temperature as a function of oxygen/ carbon feed ratio (after Yoon et al., EPRI AF-590).

spects, so quantitative conclusions should be drawn with care, but the general description of reactor characteristics appears to be correct and applicable in practice.

7.4.5 Two-Dimensional Homogeneous Model

It is relatively straightforward in principle to include radial dispersion of mass and heat in the reactor model equations. Radial convective flow is assumed to be negligible. The ordinary derivatives d/dz in Eqs. (7.7) through (7.10) now become partial derivatives $\partial/\partial z$, and dispersion terms of the form

$$\frac{1}{r}\frac{\partial}{\partial r}\left(r\frac{\partial}{\partial r}\right)$$

now appear. The *computational* problem is anything but straightforward, but

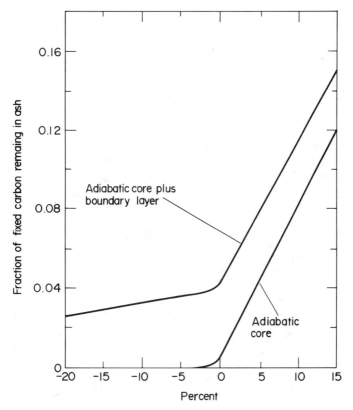

FIG. 7.12.
Unreacted carbon in discharged ash as a function of feed ratio, Illinois No. 6 coal (after Yoon et al., EPRI AF-590).

that need not be an issue here.* Some typical temperature and carbon conversion profiles are shown in Figs. 7.14 and 7.15, in this case using air as the oxidant instead of oxygen. The boundary layer is about twice the thickness estimated from the Einstein diffusion length analysis. Effluent compositions and temperatures computed using the one- and two-dimensional models are quite close to each other. There is a substantial wall effect when the throughput is reduced or a smaller reactor diameter is considered; Fig.

---

*The technique used was orthogonal collocation on finite elements, generally using three collocation points on the inner element (equivalent to the adiabatic core) and five on the outer (equivalent to the boundary layer).

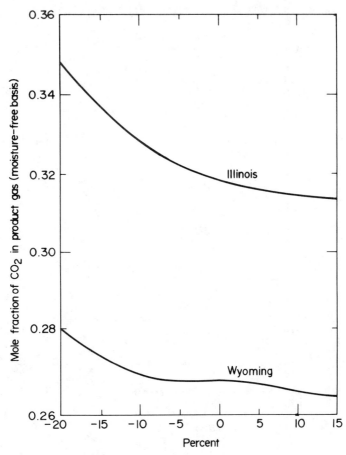

FIG. 7.13.
CO_2 in product gas (moisture free) as a function of feed conditions, Illinois No. 6 and Wyoming coals (after Yoon et al., EPRI AF-590).

7.16 shows the isotherms at 10% of the throughput in Fig. 7.14, for example, at the same relative feed conditions.

The two-dimensional model provides some useful engineering information about thermal effects that cannot be obtained reliably from the core-boundary layer formulation. Figure 7.17 shows the thermal efficiency as a function of wall heat transfer coefficient; the value of 54 BTU/ft^2-hr-°F was used in all one-dimensional simulations. A fourfold reduction in wall heat loss results only in a 1% improvement in efficiency, which is substantially

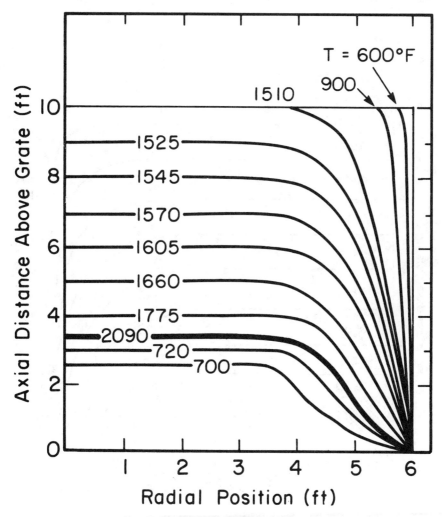

FIG. 7.14.
Isotherms, air-blown Lurgi gasifier, Illinois No. 6 coal, O_2/FC = 0.36, H_2O/O_2 = 6.7 (Denn et al., EPRI AP-2576).

offset by a one-third decrease in steam production. Even perfect insulation (and hence no steam production) results in only a 2.5% increase in efficiency. Clearly, then, wall cooling for steam production is desirable.

There could be substantial cost savings from reducing the temperature of the steam and oxidant feeds (the *blast*). Figure 7.18 shows the unreacted carbon at several radial positions as a function of blast temperature. (The adiabatic core-boundary layer formulation does do a good job here.) It is

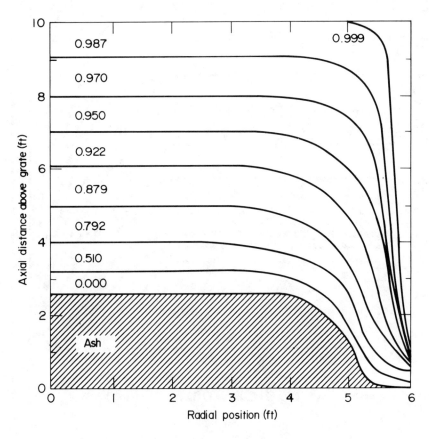

FIG. 7.15.
*Contours of unreacted carbon corresponding to Fig. 7.14
(Denn et al., EPRI AP-2576).*

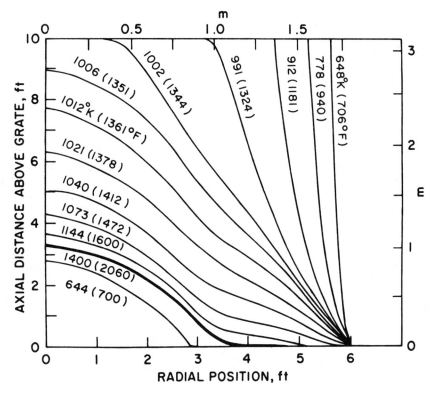

FIG. 7.16.
Isotherms at 10% carbon throughput, $O_2/FC = 0.30$, H_2O/O_2 = 6.7 (Denn et al., EPRI AP-2576).

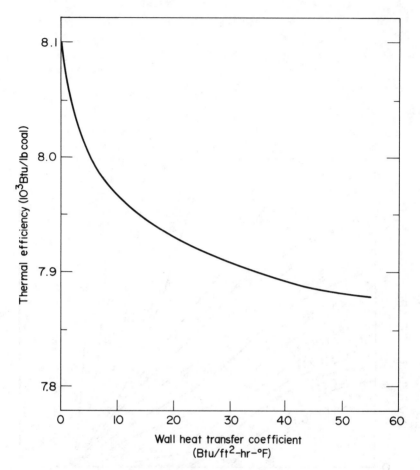

FIG. 7.17.
*Effect of wall heat-transfer coefficient on thermal efficiency
(Denn et al., EPRI AP-2576).*

clear that below about 500° F there is too much carbon in the ash, and the system cannot be operated properly; this is because the blast is not hot enough to ignite the existing coal immediately, and an induction period is required, resulting in too small a residence time for completion of the slow gasification reactions. The model is probably not a sufficiently accurate representation of the reactor to determine the critical blast temperature, but the prediction of such a phenomenon has had an impact on design and planning studies for combined-cycle power plants.

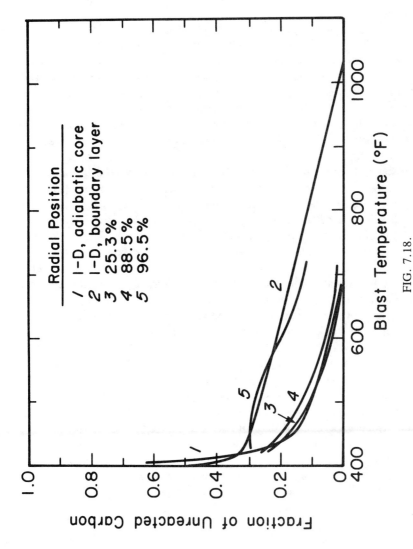

FIG. 7.18.
Fraction of unreacted carbon as a function of blast temperature (Denn et al., EPRI AP-2576).

7.4.6 Transient Homogeneous Model

It is, again, relatively straightforward to add time dependence to the equations for the homogeneous model, though the computational problem is again a very difficult one.* The transient problem can be treated at two levels, depending on the magnitude of change.

A pseudo-steady-state analysis to follow throughput changes is based on curves like Fig. 7.11, giving the position of the combustion zone as a function of carbon/oxygen ratio. Such a curve can be constructed for each throughput, and interpreted as carbon consumption rate for a given position of the maximum temperature. If the coal bed is stationary, then the position of the combustion zone will move at a thermal wave velocity corresponding to the consumption rate. The combustion zone will stabilize at a position at which the thermal wave velocity corresponds to the coal feed rate. This is equivalent to the assumption that the reactor passes through a sequence of steady states during a transient, and depends on the fact that the gas residence time is short relative to the solid residence time and the time for passage of the thermal wave. (The latter is of order hours.) This type of model has been used in evaluations of overall control strategies for gasification-combined cycle power plants.

The thermal wave analysis is not adequate for large changes associated with startup from, or turndown to, a significantly reduced throughput or a hot standby state. Solution of the full transient model equations shows that the long-term transient response of the dry ash gasifier is approximately first order, with a time constant that depends on the operational level but (based on limited calculations) appears to be independent of the magnitude of the change. Typical computed transients for an air-blown dry ash Lurgi gasifier with Illinois No. 6 coal are shown in Fig. 7.19. The time constant for turndown is about six hours, while the time constant for a load increase is about three hours. The transient time for startup is shorter than for turndown because an increase in the gas flux increases the thermal wave velocity. Published experimental data on transients in dry ash gasifiers are not available, but the long time constants are consistent with the known operating characteristics of these reactors. The dynamic model has been used to study feed policies during turndown that will maintain the combustion zone in a fixed position, with the conclusion that a step change to the new steady-state ratio is effective.

---

*Sections 7.4.5 and 7.4.6, though discussed only briefly and incompletely, are some of the major results of a number of years of work. The full details comprise the bulk of a Ph.D. dissertation. The computational method used was exponential collocation in time and orthogonal collocation in the radial direction, in order to be able to integrate through the radial- and time-dependent hot spot in the axial direction.

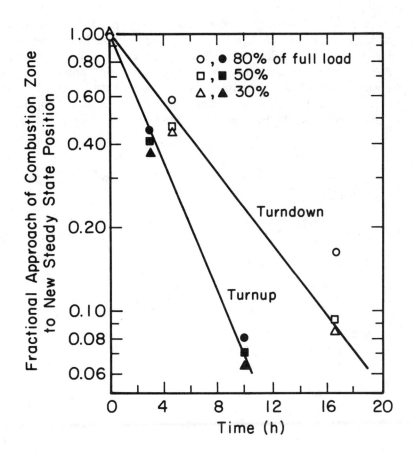

FIG. 7.19.
Computed transient response of a dry-ash Lurgi gasifier to load charges (Denn et al., EPRI AP-2576).

7.5 SENSITIVITY

As noted above, the only data available to test the gasifier model are effluent compositions and temperatures; data from within the gasifier core are unavailable, and perhaps unattainable. Sensitivity of the observable quantities to several model parameters is of interest, but one calculation suffices to establish the main point. Figure 7.20 shows the computed maximum temperature for oxygen gasification of Illinois No. 6 coal under conditions corresponding to the maximum efficiency in Fig. 7.9 as the combustion selectivity varies from 100% CO to 100% CO_2; the selectivity, as noted above, is the sole adjustable parameter in the model. The maximum temperature varies by more than 300° F, while the computed composition at the top of

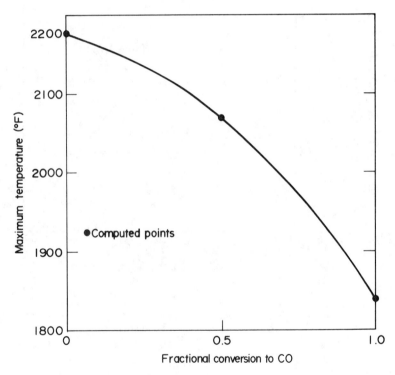

FIG. 7.20.
Sensitivity of maximum temperature to oxidation selectivity (Denn et al., 1979, copyright American Chemical Society, reproduced with permission).

7.5 SENSITIVITY

the reaction zone (before devolatilization and drying), given as the first column in Table 7.2, remains unchanged to three significant figures! It should be recalled that the maximum temperature is the major factor in operability, since it determines the state of the ash. Similar insensitivity of the effluent is observed to other factors that influence the maximum temperature and the internal profiles.

The explanation of this behavior is readily obtained from a kinetics-free analysis, Section 7.3, which for complete carbon conversion leads to the results in Table 7.2. (The results for the detailed model and the maximum temperature in the kinetics-free calculation are based on 50% CO, 50% CO_2.) Complete conversion of carbon and oxygen and equilibrium of the water-gas shift of the reaction is sufficient to define the effluent conditions *independently of phenomena occurring within the reactor*. Agreement between the detailed and kinetics-free calculations is nearly perfect for the Illinois coal; the kinetics utilized in the detailed model do allow some methanation to occur for the Wyoming coal, while methanation is not included in the kinetics-free calculation. The kinetics-free approach is clearly an effective means of estimating process performance *if* the feed conditions necessary for complete carbon conversion are known. (The maximum temperature estimated in this way will differ significantly from the detailed model calculation only if the particle size is sufficiently large that the oxidation rate is limited by mass transfer and becomes comparable to the steam gasification rate, or if large amounts of methane are formed.)

TABLE 7.2
Comparison of detailed and kinetics-free model calculations for oxygen-blown gasification of Illinois and Wyoming coals in a dry-ash Lurgi gasifier (Denn et al., 1979, Copyright American Chemical Society, reproduced with permission).

	Illinois Coal		Wyoming Coal	
	Detailed Model	Kinetics-free	Detailed Model	Kinetics-free
H_2O	0.68	0.68	0.49	0.45
H_2	0.32	0.32	0.45	0.55
CO	0.09	0.09	0.22	0.25
CO_2	0.22	0.22	0.30	0.30
CH_4	0.002	0	0.03	0
Exit temperature, °F	1441	1437	1383	1320
H_2/CO	3.43	3.52	2.07	2.21
Maximum temperature, °F	2072	2004	2361	2362

This issue of sensitivity and model validation is of the utmost importance, and we shall return to it in Chapter 13.

7.6 CONCLUDING REMARKS

The modeling case study sketched here is but one of a number of approaches taken to the analysis of this process, but it is probably the most widely used. It has been employed in several systems studies of gasification and power plants, and continues to be used at the time of this writing for similar purposes (as well as for preliminary design studies of more efficient dry-ash configurations). It therefore represents a nonproprietary look at the several levels of modeling that can be used to advantage on a complex process. The basic principles outlined in the earlier chapters are clearly applicable, and were used, though the resulting equations are considerably more complex than those encountered previously.

The issues regarding measurement and model validation that were touched upon at the end of the chapter are of primary importance. They are deferred for further discussion only because they are best discussed in the context of another case study as well.

BIBLIOGRAPHICAL NOTES

The subject of this chapter is treated in more detail, and in the context of other types of gasifiers, in

> M. M. Denn and R. Shinnar, "Coal Gasification Reactors," in J. J. Carberry and R. Varma, eds. *Chemical Reaction and Reactor Engineering* (New York, NY: Marcel Decker, 1986).

Details of the model development and numerical computation are in two reports to the Electric Power Research Institute:

> H. Yoon, J. Wei, and M. M. Denn, "Modeling and Analysis of Moving Bed Coal Gasifiers," *ERPI AF-590*, Vol. 1, Nov. 1977; Vol. 2, Feb. 1978.
>
> M. M. Denn, J. Wei, W. C. Yu, and R. Cwiklinski, "Detailed Simulation of a Moving-Bed Gasifier," *ERPI AP-2576*, Sept. 1982.

Application in a systems study of a gasification-combined-cycle power plant is in

7.6 CONCLUDING REMARKS

> D. J. Ahner, A. S. Brower, M. H. Dewes, and A. S. Patel, "Moving-Bed Gasification-Combined-Cycle Control Study," *EPRI AP-1740*, Vol. 1, March, 1981.

A recent application to explore alternative feed configurations for increasing the efficiency of dry-ash gasifiers is described in

> C. H. Lu and M. M. Denn, in *Frontiers in Chemical Reaction Engineering* (New Delhi: Wiley Eastern, 1984).

The model is described in the following journal articles:

> M. M. Denn and J. Wei, *Proc. Joint Meeting of Chemical Engineering, CIESC/AIChE* (New York, NY: Am. Inst. Chem. Engrs., 1982) p. 631.
>
> M. M. Denn, W. C. Yu, and J. Wei, *Ind. Eng. Chem. Fundamentals* **18** (1979) 286.
>
> H. Yoon, J. Wei, and M. M. Denn, *AIChE J.* **24** (1978) 885.
>
> H. Yoon, J. Wei, and M. M. Denn, *Ind. Eng. Chem. Process Des. and Dev.* **18** (1979) 306.
>
> H. Yoon, J. Wei, and M. M. Denn, *AIChE J.* **25** (1979) 429.
>
> W. C. Yu, M. M. Denn, and J. Wei, *ACS Symposium Series 196* (1982) 359.
>
> W. C. Yu, M. M. Denn, and J. Wei, *Chem. Eng. Sci.* **38** (1983) 1467.

8

STABILITY

8.1 INTRODUCTION

Engineering systems are usually designed to operate in the steady state, but they are always subject to input disturbances. There are two fundamentally different classes of response to input disturbances that have sometimes been confused in the engineering literature. One is the usual stable transient that is often illustrated by means of a Bode diagram like Fig. 1.2; the system always operates in the neighborhood of the design steady state and returns to the steady state in the absence of disturbances, but input disturbances of various frequencies might be amplified or attenuated as they propagate through the process. Such a stable system might be very sensitive dynamically if there is a frequency range where the amplitude ratio is large.

An inherently unstable system is one in which the steady state cannot be maintained in the presence of *any* input disturbance, no matter how small. (It is not always an easy matter to distinguish strong sensitivity to very small disturbances from inherent instability, although the latter will often show a more regular transient structure.) The instability of the steady state can sometimes be inferred from the steady-state model equations themselves, as shown in Sections 4.5 and 5.7, but an analysis of the full transient model is more commonly required. The ability to predict transients such as the onset of an instability is a sensitive test of a process model.

There are several analytical approaches to studying the stability of a steady state, resulting in part from different questions that can be asked. We might wish to know the range of process parameters for which we can demonstrate that the system is inherently unstable; this is, in fact, the easiest stability question to treat, and the one that dominates the applications literature. We might wish to establish the process conditions under which the system is stable regardless of the shape or magnitude of the input disturbances; this is a far more difficult problem because of the generality of the result that is sought. Somewhat less general, but similar analyses can be carried out to establish the disturbance magnitude for which stability can be ensured and the disturbance magnitude at which a stable steady state must become unstable. We shall touch briefly on all of the approaches except the last in this chapter.

8.2 LIAPUNOV FUNCTION

We shall suppose that the process model consists of a set of ordinary differential equations,

$$\frac{dx_1}{dt} = \dot{x}_1 = f_1(x_1, x_2, x_3, \ldots, x_n) \tag{8.1a}$$

$$\frac{dx_2}{dt} = \dot{x}_2 = f_2(x_1, x_2, x_3, \ldots, x_n) \tag{8.1b}$$

$$\vdots$$

$$\frac{dx_n}{dt} = \dot{x}_n = f_n(x_1, x_2, x_3, \ldots, x_n) \tag{8.1c}$$

where $\{x_i\}$ are the characterizing variables and $\{f_i\}$ are functions that depend on the $\{x_i\}$ and some process parameters, but are independent of time. If we think of the $\{x_i\}$ as components of an n-vector, $\mathbf{x}$, and the f_i as components of a vector function $\mathbf{f}(\mathbf{x})$, then we can write Eq. (8.1) in the compact form

$$\frac{d\mathbf{x}}{dt} = \dot{\mathbf{x}} = \mathbf{f}(\mathbf{x}) \tag{8.2}$$

The steady states, denoted $\mathbf{x}_s$, are solutions of the n simultaneous algebraic equations

$$\mathbf{f}(\mathbf{x}_s) = \mathbf{0} \tag{8.3}$$

We now need to define a norm, or measure of distance of a solution $\mathbf{x}(t)$ from a steady state $\mathbf{x}_s$. Let $\boldsymbol{\xi}$ represent the deviation from the steady state,

$$\xi(t) \equiv \mathbf{x}(t) - \mathbf{x}_s \tag{8.4}$$

Then $\xi(t)$ satisfies the vector differential equation

$$\dot{\xi} = \mathbf{f}(\mathbf{x}_s + \xi) \tag{8.5}$$

Any positive definite scalar function* $V(\xi)$ can be a suitable measure of distance from the steady state. We now consider any trajectory in time, $\xi(t)$, starting from some initial nonzero value ξ_0. If $V(\xi(t))$ is smaller than $V(\xi_0)$, then the system is closer to the steady state at time t than at time zero. It then follows that, if $dV(\xi(t))/dt$ is always negative, V is a continuously decreasing function of time and the process is always coming closer to the origin $\xi = \mathbf{0}$, which must ultimately be reached. This is clearly a criterion for asymptotic stability, in that the system returns to the steady state following a displacement.

We can now formalize this development somewhat. Suppose we have a simply connected region containing the origin in which there exists a function $V(\xi)$ which is positive definite, and in which the derivative $dV(\xi)/dt$ is negative definite. All solutions of Eq. (8.5) having initial conditions ξ_0 within the region must then approach the origin, $\xi = \mathbf{0}$, and the system is asymptotically stable with respect to disturbances that create initial conditions within this region. The existence of such a function $V(\xi)$, which is called a *Liapunov function*, is therefore a sufficient condition for asymptotic stability. Reversal of the arguments readily demonstrates that the steady state must be unstable if both $V(\xi)$ and dV/dt are positive definite within a simply connected region containing the origin.

It is important to note that this proof is not constructive. It simply tells us what must occur if a Liapunov function exists. It does not tell us that one exists, or does not exist, or how to find one. The possible advantage is that the approach is strictly algebraic, as the following application of the chain rule shows:

$$\frac{dV(\xi(t))}{dt} = \sum_{i=1}^{n} \frac{\partial V}{\partial \xi_i} \dot{\xi}_i = \sum_{i=1}^{n} \frac{\partial V}{\partial \xi_i} f_i = \mathbf{f} \cdot \nabla V \tag{8.6}$$

Here, ∇V denotes the gradient with respect to ξ, and $\mathbf{f} \cdot \nabla V$ is the vector inner (dot) product. We therefore reach conclusions about the stability of the dynamical system defined by Eq. (8.1) or (8.5) by examining the properties of $V(\xi)$ and $\mathbf{f}(\mathbf{x}_s + \xi) \cdot V(\xi)$ in a region of ξ-space that includes the origin. The differential equations themselves need never be solved.

---

*That is, $V(\xi) > 0$ if $\xi \neq \mathbf{0}$; $V(\mathbf{0}) = 0$.

8.2 LIAPUNOV FUNCTION

The mass-spring-dashpot system in Section 1.2 provides a nice illustrative example. If we take the equilibrium position to be $x_s = 0$, then the governing equation in the absence of external forcing for an arbitrary spring and damping function can be written

$$\ddot{\xi} + \mu(\xi,\dot{\xi})\,\dot{\xi} - F_s(\xi) = 0 \qquad (8.7)$$

where F_s is expressible in terms of a positive definite potential, $\Phi(\xi)$:

$$F_s = -\frac{d\Phi(\xi)}{d\xi} \qquad (8.8)$$

Equivalently, we may define new variables ξ_1 and ξ_2 as follows:

$$\xi \equiv \xi_1, \qquad \dot{\xi} \equiv \xi_2 \qquad (8.9)$$

Equation (8.7) is then equivalent to the pair of first-order equations

$$\dot{\xi}_1 = \xi_2 \qquad \equiv f_1(\xi_1,\xi_2) \qquad (8.10a)$$

$$\dot{\xi}_2 = -\mu(\xi_1,\xi_2)\,\xi_2 - \frac{d\Phi(\xi_1)}{d\xi_1} \equiv f_2(\xi_1,\xi_2) \qquad (8.10b)$$

We choose the sum of potential and kinetic energies as the Liapunov function:

$$V(\xi_1,\xi_2) = \Phi(\xi_1) + \tfrac{1}{2}\xi_2^2 \qquad (8.11)$$

The components of ∇V are then $d\Phi/d\xi_1$ and ξ_2, and we have

$$\frac{dV}{dt} = \mathbf{f}\cdot\nabla V = \frac{d\Phi}{d\xi_1}\xi_2 + \xi_2\left(-\mu(\xi_1,\xi_2)\xi_2 - \frac{d\Phi}{d\xi_1}\right) = -\mu(\xi_1,\xi_2)\xi_2^2 \qquad (8.12)$$

dV/dt is negative definite, and the equilibrium is stable, if and only if $\mu(\xi_1,\xi_2)$ is a positive definite function. Since μ is proportional to the viscosity of the dashpot fluid, this is simply a statement that the viscosity must be strictly positive for stability of the equilibrium. The physical meaning is obvious: a positive viscosity corresponds to dissipation of mechanical energy to heat the fluid. A negative viscosity would mean that heat was extracted from the dashpot fluid and converted to mechanical energy, which is a violation of the second law of thermodynamics.

There are physical systems that are mathematically identical to the mass-spring-dashpot system for which the coefficient $\mu(\xi,\dot\xi)$ is negative in a region around the origin. One such system is a linear oscillating electrical circuit with resistance coupled inductively to a triode, which is modeled by the *Van der Pol equation*, with $\Phi(\xi) = \frac{1}{2}\xi^2$ and $\mu = \varepsilon(\xi^2 - 1)$, $\varepsilon > 0$. This system exhibits sustained oscillations, and the equilibrium $\xi = 0$ is unstable.

8.3 STABILITY TO INFINITESIMAL PERTURBATIONS

It is rarely possible to deal with the algebraic problem described in the preceding section for arbitrary process models. A completely general result can be obtained, however, if we are willing to restrict ourselves to infinitesimal disturbances about the steady state. If the vector function **f** is twice differentiable, then we can expand the right-hand side of Eq. (8.5) in a Taylor series to obtain

$$\dot{\xi} = \mathbf{A} \cdot \xi + \mathbf{o}(\xi) \tag{8.13}$$

where*

$$\mathbf{A} = \nabla \mathbf{f}(\mathbf{x}_s); \qquad A_{ij} = \frac{\partial f_i(\mathbf{x}_s)}{\partial x_j} \tag{8.14}$$

and $\mathbf{o}(\xi)$ represents terms that go to zero faster than $|\xi|$:

$$\lim_{|\xi| \to 0} \frac{\mathbf{o}(\xi)}{|\xi|} = \mathbf{0} \tag{8.15}$$

We take a quadratic form as our positive-definite function:

$$V = \tfrac{1}{2}\xi \cdot \mathbf{B} \cdot \xi \tag{8.16}$$

where the matrix **B** is positive definite; that is, $\mathbf{y} \cdot \mathbf{B} \cdot \mathbf{y}$ is strictly positive for arbitrary nonzero vectors **y**. There is no loss in generality in taking **B** to be symmetric ($\mathbf{B}^\dagger = \mathbf{B}$), since only the symmetric part of a matrix enters

---

*We will use the inner, or dot, product of vector analysis for matrix multiplication:

$$(\mathbf{A} \cdot \xi)_i = \sum_{j=1}^{n} A_{ij}\xi_j, \qquad (\xi \cdot \mathbf{A})_j = \sum_{i=1}^{n} \xi_i A_{ij}$$

into the terms in a quadratic form. The condition that dV/dt be negative is then

$$\frac{dV}{dt} = \boldsymbol{\xi} \cdot (\mathbf{B} \cdot \mathbf{A}) \cdot \boldsymbol{\xi} + \boldsymbol{\xi} \cdot \mathbf{B} \cdot \mathbf{o}(\boldsymbol{\xi}) < 0 \qquad (8.17)$$

It is convenient to define a new normalized variable, $\mathbf{y} = \boldsymbol{\xi}/|\boldsymbol{\xi}|$, and to divide Eq. (8.17) by $|\boldsymbol{\xi}|^2$, to obtain

$$\mathbf{y} \cdot (\mathbf{B} \cdot \mathbf{A}) \cdot \mathbf{y} + \mathbf{y} \cdot \mathbf{B} \cdot \left(\frac{\mathbf{o}(\boldsymbol{\xi})}{|\boldsymbol{\xi}|}\right) < 0 \qquad (8.18)$$

If we now take the limit as $|\boldsymbol{\xi}| \to 0$ (i.e., an *infinitesimal perturbation*) the second term vanishes, since the limit of $\mathbf{o}(\boldsymbol{\xi})/|\boldsymbol{\xi}|$ is zero, and we have

$$|\boldsymbol{\xi}| \to 0: \quad \mathbf{y} \cdot (\mathbf{B} \cdot \mathbf{A}) \cdot \mathbf{y} < 0 \qquad (8.19)$$

It therefore follows that the function $V(\boldsymbol{\xi})$ defined by Eq. (8.16) is a Liapunov function for infinitesimal perturbations, and the system described by Eq. (8.13) is asymptotically stable, if the matrix $\mathbf{B} \cdot \mathbf{A}$ is negative definite. While we shall not present a proof here, it is readily shown that $\mathbf{B} \cdot \mathbf{A}$ *is negative definite for symmetric positive definite* $\mathbf{B}$ *if and only if all eigenvalues of* $\mathbf{A}$ *have negative real parts*.

The preceding steps can be repeated to show that there is always some infinitesimal disturbance that grows and allows dV/dt to be positive if any eigenvalue of $\mathbf{A}$ has a positive real part. Thus, one eigenvalue with a positive real part is a sufficient condition for instability, and all eigenvalues having negative real parts is a sufficient condition for stability. *The necessary and sufficient condition for asymptotic stability to infinitesimal perturbations is therefore that all eigenvalues of* $\mathbf{A} = \nabla \mathbf{f}(\mathbf{x}_s)$ *have negative real parts*.

8.4 CONTINUOUS-FLOW STIRRED REACTOR: I

The equations describing the continuous-flow stirred chemical reactor, with a single reaction and rapid dynamics in the cooling jacket, were derived in Section 5.7. This system can exhibit three steady-state solutions for even the simplest kinetic constitutive equation.

Equations (5.40) and (5.41) are written

$$\dot{x}_1 = 1 - x_1 - \alpha x_1 e^{-\gamma/x_2} = f_1(x_1, x_2) \qquad (8.20a)$$

$$\dot{x}_2 = (1 + \delta)[\phi - x_2 + \alpha\beta x_1 e^{-\gamma/x_2}] = f_2(x_1, x_2) \qquad (8.20b)$$

where x_1 and x_2 are dimensionless concentration and temperature, respectively. (x_1 and x_2 are denoted x and y in Chapter 5.) Time is made dimensionless with respect to the mean residence time, V/q. The parameters are defined as follows:

$$\alpha = \frac{k_0 V}{q}, \qquad \delta = \frac{hA}{\rho q c_p(1 + \mathcal{H})}, \qquad \beta = \frac{(-\Delta \underset{\sim}{H}_R)c_{Af}}{\rho c_p T_f(1 + \delta)},$$

$$\gamma = \frac{E}{RT_f}, \qquad \phi = \frac{1 + \delta T_{jf}/T_f}{1 + \delta} \qquad (8.21)$$

$\delta = 0$ corresponds to an adiabatic system, with no heat transfer to the surroundings. We established in Section 5.7 that the middle steady state must be unstable because the *slope condition*

$$\alpha F'(x_{2s}) < 1 \qquad (8.22)$$

is not satisfied, where

$$F(x_2) = (\beta + \phi - x_2)e^{-\gamma/x_2} \qquad (8.23)$$

but we were unable to make any statements about the other steady states.

The matrix $\mathbf{A} = \nabla \mathbf{f}$ has elements

$$A_{11} = \frac{\partial f_1}{\partial x_1} = -1 - \alpha e^{-\gamma/x_{1s}} \qquad (8.24a)$$

$$A_{12} = \frac{\partial f_1}{\partial x_2} = -\frac{\alpha \gamma}{\beta x_{2s}^2} F(x_{2s}) \qquad (8.24b)$$

$$A_{21} = \frac{\partial f_2}{\partial x_1} = (1 + \delta)\alpha\beta e^{-\gamma/x_{2s}} \qquad (8.24c)$$

$$A_{22} = \frac{\partial f_2}{\partial x_2} = -(1 + \delta)(1 + \beta A_{12}) \qquad (8.24d)$$

The eigenvalues λ_1 and λ_2 are solutions of the characteristic equation

$$\lambda^2 - (A_{11} + A_{12})\lambda + (A_{11}A_{22} - A_{12}A_{21}) = 0 \qquad (8.25)$$

and the necessary and sufficient conditions that the roots have negative real parts are

$$A_{11}A_{22} - A_{12}A_{21} > 0 \qquad (8.26a)$$

$$A_{11} + A_{22} < 0 \tag{8.26b}$$

Equations (8.26) are equivalent to the following two inequalities:

$$\alpha F'(x_{2s}) < 1 \tag{8.27a}$$

$$\alpha F'(x_{2s}) < 2 + \delta \left[1 - \frac{\alpha \gamma}{x_{2s}^2} F(x_{2s}) \right] \tag{8.27b}$$

These inequalities are necessary and sufficient conditions for a steady state to be stable to infinitesimal perturbations. For an adiabatic reactor ($\delta = 0$) the second inequality must always be satisfied when the first is satisfied. Equation (8.27a), which is the slope condition, is equivalent to a positive spring constant in the mass-spring-dashpot analogy, while Eq. (8.27b) is equivalent to a positive viscous damping coefficient.

It follows from Eq. (8.27b) that a steady state design for a nonadiabatic reactor might be unique and satisfy the slope condition, and yet be unstable to infinitesimal perturbations and hence unattainable in practice. The data in Fig. 5.4 are for precisely such a reacting system. Since there is only one steady state, and it is unstable, the system cycles continuously, even though feed conditions are constant.

8.5 CONTINUOUS-FLOW STIRRED REACTOR: II

The linear analysis in the preceding section is very powerful, and establishes the conditions under which the reactor will become unstable to any disturbance. It gives no information, however, about the extent of the stable region for cases in which the steady state is stable to infinitesimal disturbances. This problem is far more difficult, since it requires construction of a Liapunov function that can be used in conjunction with the full nonlinear reactor equations. There are few general techniques for finding such Liapunov functions, and hence the approach is of limited utility. Some useful results can be obtained for special cases, however, including the flow reactor, which we shall again use for illustration.

Equations (8.20) are written in terms of deviation from the steady state as

$$\dot{\xi}_1 = -\xi_1 - \alpha x_{1s} \Delta(x_2, x_{2s}) \xi_2 - \alpha \xi \exp(-\gamma/x_2) \tag{8.28a}$$

$$\dot{\xi}_2 = (1 + \delta)[-\xi_2 + \alpha \beta x_{1s} \Delta(x_2, x_{2s}) \xi_2 + \alpha \beta \xi \exp(-\gamma/x_2)] \tag{8.28b}$$

We have sometimes used x_2 in place of $x_{2s} + \xi_2$ in order to simplify some of the nomenclature. The function $\Delta(x_2, x_{2s})$ is defined

$$\Delta(x_2, x_{2s}) = \frac{1}{\xi_2}\left[\exp\left(-\frac{\gamma}{x_{2s} + \xi_2}\right) - \exp\left(-\frac{\gamma}{x_{2s}}\right)\right] \tag{8.29}$$

Note that

$$\lim_{\xi_2 \to 0} \Delta(x_2, x_{2s}) = \frac{\gamma}{x_{2s}^2} \exp(-\gamma/x_{2s}) \tag{8.30}$$

The simplest positive-definite function is a sum of squares,

$$V = \frac{1}{2}(\xi_1^2 + \xi_2^2) \tag{8.31}$$

(In the fluid mechanics literature, the use of a sum-of-squares Liapunov function is often called the *energy method*.) We then have

$$\begin{aligned}\frac{dV}{dt} &= \xi_1 \dot{\xi}_1 + \xi_2 \dot{\xi}_2 \\ &= -(1 + \alpha a^{-\gamma/x_2})\xi_1^2 \\ &\quad - \alpha[x_{1s}\Delta(x_2, x_{2s}) - (1 + \delta)\beta e^{-\gamma/x_2}]\xi_1\xi_2 \\ &\quad - (1 + \delta)[1 - \alpha\beta x_{1s}\Delta(x_2, x_{2s})\xi_2^2]\end{aligned} \tag{8.32}$$

dV/dt is expressed here as a quadratic form in ξ_1 and ξ_2. The necessary and sufficient conditions that the quadratic form be negative definite are

$$1 + \alpha e^{-\gamma/x_2} > 0 \tag{8.33a}$$

$$4(1 + \delta)(1 + \alpha e^{-\gamma/x_2})[1 - \alpha\beta x_{1s}\Delta(x_2, x_{2s})]$$
$$- \alpha^2[x_{1s}\Delta(x_2, x_{2s}) - (1 + \delta)\beta e^{-\gamma/x_2}]^2 > 0 \tag{8.33b}$$

The first of these inequalities is always satisfied, so negative definiteness depends only on the second, which is a function only of x_2. Let x_{2c} be the first value of x_2 for which inequality (8.33b) is violated. Stability is then ensured for all initial conditions in the region

$$(x_1 - x_{1s})^2 + (x_2 - x_{2s})^2 < (x_{2c} - x_{2s})^2 \tag{8.34}$$

Larger deviations from the steady state might (and probably will) be stable,

8.5 CONTINUOUS FLOW STIRRED REACTOR: II

but we are unable to establish this fact with the particular choice of Liapunov function.

Inequality (8.33b) is conservative, and it is possible that it will not be satisfied at a stable steady state. This is most easily demonstrated for an adiabatic reactor, $\delta = 0$. In the limit $x_2 \to x_{2s}$ we can use Eq. (8.30) and rewrite the inequality as

$$\delta = 0, \text{ limit:} \quad \frac{1 - \alpha\beta\gamma x_{1s}e^{-\gamma/x_{2s}}}{x_{2s}^2} > \frac{\alpha^2 e^{-2\gamma/x_{2s}}[x_{1s}\gamma/x_{2s})^2 - \beta]^2}{4[1 + \alpha e^{-\gamma/x_{2s}}]} \quad (8.35)$$

On the other hand, the slope condition, Eq. (8.27a), is

$$1 - \frac{\alpha\beta\gamma \, x_{1s}e^{-\gamma/x_{2s}}}{x_{2s}^2} > -\alpha e^{-\gamma/x_{2s}} \quad (8.36)$$

Thus, the slope condition can be satisfied by an adiabatic reactor, establishing asymptotic stability to infinitesimal disturbances, and yet the sum-of-squares Liapunov function might not exist.

The following example illustrates this last point. Consider an adiabatic reactor ($\delta = 0$) with the following parameters:

$$\alpha = 1.8 \times 10^2, \quad \beta = 0.554, \quad \gamma = 29.9$$

There are steady states that are stable according to the slope condition at

$$x_{1s} = 0.014, \quad x_{2s} = 1.546 \text{ (high temperature)}$$

$$x_{1s} = 0.998, \quad x_{2s} = 1.001 \text{ (low temperature)}$$

and an unstable intermediate steady state at

$$x_{1s} = 0.532, \quad x_{2s} = 1.259$$

Inequality (8.33b) establishes a stability region

$$[(x_1 - 0.998)^2 + (x_2 - 1.001)^2]^{1/2} < 0.197$$

about the low-temperature steady state. This is a conservative result; for initial values of dimensionless temperature x_2 less than about 1.2, the low-temperature steady state is stable for all dimensionless concentrations. Thus, the real stability limit is defined quite closely for temperature perturbations accompanied by essentially no concentration perturbation ($x_1 \sim 1.0$), but is poorly represented for finite concentration perturbations.

Inequality (8.33b) is not satisfied by any value of x_2 at the high-temperature steady state; inequality (8.35) is not satisfied, although the slope condition is. This steady state is stable to all concentration perturbations, and to all temperature perturbations down to a dimensionless value of about 1.35. Thus, a sum-of-squares Liapunov function gives no information in this case about stability.

We could have obtained a finite stability region about the high-temperature steady state, and a larger stability region about the low-temperature steady state, by using a more general positive-definite quadratic form for the Liapunov function,

$$V = \alpha_{11}\xi_1^2 + 2\alpha_{12}\xi_1\xi_2 + \alpha_{22}\xi_2^2$$

$$\alpha_{11} > 0, \qquad \alpha_{11}\alpha_{22} - \alpha_{12}^2 > 0$$

and searching for the values of the parameters α_{ij} giving the largest stability region. The essential point that we wished to make here, however, in addition to illustrating the method, is that this approach to stability of nonlinear processes is extremely limited. Yet, it is the only approach available except for direct numerical simulation. Special techniques do exist to construct Liapunov functions for certain classes of problems, but the results are almost always very conservative. The following function is a Liapunov function for the adiabatic reactor equations, for example:

$$V = \frac{1}{2}\left[\xi_1 + \frac{1}{\alpha\beta}\int_0^{\xi_2} e^{\gamma/(x_{2s}+\xi_2)}\,d\xi_2 + \frac{1}{\beta}\xi_2\right]^2$$
$$+ \int_0^{\xi_2} \frac{[1 + \alpha e^{-\gamma/x_2} - \alpha\beta x_{1s}\Delta(x_2,x_{2s})]\xi_2}{[\alpha\beta e^{-\gamma/x_2}]^2}\,d\xi_2 \qquad (8.37)$$

It can be shown to reduce to the slope condition as the steady state is approached, and hence to give finite stability regions about any stable steady state, but these regions are again small.

8.6 FINITE STABILITY REGIONS FROM LINEAR THEORY

There is often a close connection between problems in stability and problems in optimization. This link is clearly illustrated in the attempt to obtain the largest stability region for a Liapunov function of a particular form. We shall illustrate an analytical approach to that problem here, both for its own inherent interest and because it yields a potentially useful result: *an estimate (albeit conservative) of a region of stability to finite disturbances that re-*

8.6 FINITE STABILITY REGIONS FROM LINEAR THEORY

quires only the solution of a linear eigenvalue problem. For systems of high order, including distributed-parameter (infinite-order) processes, a linear eigenvalue problem can be solved rather easily, while a systematic study in the high- (or infinite-) dimensional space to construct the boundaries of the stability region would be difficult or impossible.

We will consider a system of ordinary differential equations of the form

$$\dot{\boldsymbol{\xi}} = -\mathbf{D} \cdot \boldsymbol{\xi} + \alpha \mathbf{E}(\boldsymbol{\xi}) \cdot \boldsymbol{\xi} \tag{8.38a}$$

$$\dot{\xi}_i = -\sum_j D_{ij}\xi_j + \alpha \sum_j E_{ij}\xi_j \tag{8.38b}$$

$\alpha > 0$ is a process parameter. $\mathbf{D}$ is taken to be positive definitive, so the system is asymptotically stable as $\alpha \to 0$. $\mathbf{E}$ is a function of $\boldsymbol{\xi}$. (The stirred-tank reactor equations are of this form.) We seek the largest value of α for which stability can be ensured.

The approach is conveniently illustrated using a sum-of-squares Liapunov function, although it will be evident that it is more general than shown here. Take

$$V = \frac{1}{2}\boldsymbol{\xi} \cdot \boldsymbol{\xi} \tag{8.39}$$

Then we have asymptotic stability if

$$\frac{dV}{dt} = -\boldsymbol{\xi} \cdot \mathbf{D} \cdot \boldsymbol{\xi} + \alpha \boldsymbol{\xi} \cdot \mathbf{E}(\boldsymbol{\xi}) \cdot \boldsymbol{\xi} < 0 \tag{8.40}$$

Since α and $\boldsymbol{\xi} \cdot \mathbf{D} \cdot \boldsymbol{\xi}$ are both positive, we may divide by both and retain the sense of the inequality:

$$\frac{1}{\alpha} > \frac{\boldsymbol{\xi} \cdot \mathbf{E}(\boldsymbol{\xi}) \cdot \boldsymbol{\xi}}{\boldsymbol{\xi} \cdot \mathbf{D} \cdot \boldsymbol{\xi}} \tag{8.41}$$

The inequality is trivially satisfied if $\mathbf{E}(\boldsymbol{\xi})$ is negative definite, in which case the system is asymptotically stable for all $\alpha > 0$.

If $\mathbf{E}(\boldsymbol{\xi})$ is not negative definite, then $1/\alpha$ must exceed the right-hand side of Eq. (8.41) for stability. We define the maximum over all $\boldsymbol{\xi}$ of $\boldsymbol{\xi} \cdot \mathbf{E} \cdot \boldsymbol{\xi}/\boldsymbol{\xi} \cdot \mathbf{D} \cdot \boldsymbol{\xi}$ as $1/\lambda$ and write

$$\frac{1}{\alpha} > \frac{1}{\lambda} = \max_{\boldsymbol{\xi}} \frac{\boldsymbol{\xi} \cdot \mathbf{E}(\boldsymbol{\xi}) \cdot \boldsymbol{\xi}}{\boldsymbol{\xi} \cdot \mathbf{D} \cdot \boldsymbol{\xi}} \tag{8.42}$$

The maximum is obtained by setting derivatives with respect to components of $\boldsymbol{\xi}$ to zero, leading to the following equation:

$$\sum_j (D_{ij} + D_{ji}) \xi_j^* = \lambda \left[\sum_j (E_{ij} + E_{ji}) \xi_j^* + \sum_{j,k} \xi_j^* \frac{\partial E_{jk}}{\partial \xi_i} \xi_k^* \right] \quad (8.43)$$

$\boldsymbol{\xi}^*$ is the maximizing value of $\boldsymbol{\xi}$. Equation (8.43) is a nonlinear eigenvalue equation which always admits the trivial solution $\boldsymbol{\xi}^* = \mathbf{0}$. If there are no positive eigenvalues, λ, then the maximum in Eq. (8.42) is always negative and stability is ensured for all values of $\boldsymbol{\xi}$. If there is at least one positive eigenvalue, then the critical value of α is equal to the smallest positive eigenvalue.

In the limit $|\boldsymbol{\xi}^*| \to 0$ the quadratic terms in Eq. (8.43) can be neglected, and the problem reduces to a linear eigenvalue problem,

$$(\mathbf{D} + \mathbf{D}^\dagger) \cdot \boldsymbol{\xi}^* = \lambda [\mathbf{E}(\mathbf{0}) + \mathbf{E}(\mathbf{0})^\dagger] \cdot \boldsymbol{\xi}^* \quad (8.44)$$

Only the symmetric parts of the matrices $\mathbf{D}$ and $\mathbf{E}(\mathbf{0})$ enter here, so the eigenvalues are known to be real. Equation (8.44) is the symmetric part of the eigenvalue equation that results from linearizing Eq. (8.38) and examining the dynamical response of the linear system. We will denote the eigenvalue to Eq. (8.44) as $\lambda^{(0)}$. The eigenvector, $\boldsymbol{\xi}^{*(0)}$, is of arbitrary magnitude, and it is convenient to take it as scaled to unity,

$$|\boldsymbol{\xi}^{*(0)}|^2 = \boldsymbol{\xi}^{*(0)} \cdot \boldsymbol{\xi}^{*(0)} = 1 \quad (8.45)$$

We now seek a series solution to Eq. (8.43) in the form

$$\boldsymbol{\xi}^* = \mathcal{A}\boldsymbol{\xi}^{*(0)} + \mathcal{A}^2\boldsymbol{\xi}^{*(1)} + \cdots \quad (8.46a)$$

$$\lambda = \lambda^{(0)} + \mathcal{A}\lambda^{(1)} + \cdots \quad (8.46b)$$

where $\mathcal{A}$ is a scalar amplitude. Substituting into Eq. (8.43) and collecting terms of the same power of $\mathcal{A}$ leads to Eq. (8.44) for $\lambda^{(0)}, \boldsymbol{\xi}^{*(0)}$, and to the following equation for $\lambda^{(1)}, \boldsymbol{\xi}^{*(1)}$:

$$\sum_j (D_{ij} + D_{ji}) \xi_j^{*(1)} = \lambda^{(0)} \sum_j [E_{ij}(\mathbf{0}) + E_{ji}(\mathbf{0})] \xi_j^{*(1)}$$
$$+ \left\{ \lambda^{(1)} \sum_j [E_{ij}(\mathbf{0}) + E_{ji}(\mathbf{0})] \xi_j^{*(0)} \right.$$
$$\left. + \lambda^{(0)} \sum_{j,k} [\gamma_{ijk} + \gamma_{jki} + \gamma_{kij}] \xi_k^{*(0)} \xi_j^{*(0)} \right\} \quad (8.47)$$

8.6 FINITE STABILITY REGIONS FROM LINEAR THEORY

$$\gamma_{ijk} = \frac{\partial E_{jk}}{\partial \xi_i} \quad \text{at} \quad \boldsymbol{\xi} = \mathbf{0} \tag{8.48}$$

The homogeneous part of Eq. (8.47) is identical to Eq. (8.44), indicating that $\boldsymbol{\xi}^{*(0)}$ is an eigenvector. Since the homogeneous part has an eigensolution, there will be a solution to the nonhomogeneous equation if and only if the nonhomogeneous part is orthogonal to the eigenvector; i.e.,

$$\lambda^{(1)} \sum_{i,j} [E_{ij}(0) + E_{ji}(0)] \xi_i^{*(0)} \xi_j^{*(0)}$$

$$+ \lambda^{(0)} \sum_{i,j,k} [\gamma_{ijk} + \gamma_{jki} + \gamma_{kij}] \xi_i^{*(0)} \xi_j^{*(0)} \xi_k^{*(0)} = 0 \tag{8.49}$$

The second sum is symmetric in i,j,k, so the expression for λ can then be written

$$\lambda = \lambda^{(0)} \left\{ 1 - \frac{3 \sum_{i,j,k} \gamma_{ijk} \xi_i^{*(0)} \xi_j^{*(0)} \xi_k^{*(0)}}{\sum_{i,j} [E_{ij}(0) + E_{ji}(0)] \xi_i^{*(0)} \xi_j^{*(0)}} \mathcal{A} + o(\mathcal{A}) \right\} \tag{8.50}$$

According to Eq. (8.42), $\lambda > \alpha$ for stability. For given α we can rearrange the inquality to find the maximum amplitude for which stability is ensured. For small $\mathcal{A}$, neglecting $o(\mathcal{A})$, we thus obtain

$$\mathcal{A} < \left| \frac{\sum_{i,j} [E_{ij}(0) + E_{ji}(0)] \xi_i^{*(0)} \xi_j^{*(0)}}{\sum_{i,j,k} \gamma_{ijk} \xi_i^{*(0)} \xi_j^{*(0)} \xi_k^{*(0)}} \right| \tag{8.51}$$

The absolute value sign is required because $\boldsymbol{\xi}^{*(0)}$ is only defined to within a 180° rotation. Equation (8.51) thus defines tne limiting radius for which stability is ensured according to the quadratic Liapunov function $V = \frac{1}{2} \boldsymbol{\xi} \cdot \boldsymbol{\xi}$. We therefore obtain an estimate of the size of the stability region by solution of a *linear* eigenvalue problem.

For the adiabatic reactor, Eq. (8.28) with $\delta = 0$, we have

$$\mathbf{D} = \begin{pmatrix} 1 & 0 \\ 0 & 1 \end{pmatrix} \tag{8.52a}$$

$$E(0) = e^{-\gamma/x_{2s}} \begin{pmatrix} -1 & \dfrac{-\gamma x_{1s}}{x_{2s}^2} \\ \beta & \dfrac{\beta x_{1s}\gamma}{x_{2s}^2} \end{pmatrix} \tag{8.52b}$$

$$\gamma_{1jk} = 0 \tag{8.52c}$$

$$\gamma_{2jk} = e^{-\gamma/x_{2s}} \begin{bmatrix} \dfrac{-\gamma}{x_{2s}^2} & \dfrac{-\gamma x_{1s}}{2x_{2s}^3}\left(\dfrac{\gamma}{x_{2s}} - 2\right) \\ \dfrac{\beta\gamma}{x_{2s}^2} & \dfrac{\beta x_{1s}\gamma}{2x_{2s}^3}\left(\dfrac{\gamma}{x_{2s}} - 2\right) \end{bmatrix} \tag{8.52d}$$

The linear estimate of a stability boundary is then obtained by solving the linear eigenvalue problem, Eq. (8.44), together with the requirement that $\lambda > \alpha$:

$$\alpha < \lambda^{(0)} = \frac{1}{F'(x_{2s})} \frac{2}{[1 + (\beta + \gamma x_{1s}/x_{2s}^2)^2/(1 - \beta\gamma x_{1s}/x_{2s})^2]^{1/2}} \tag{8.53}$$

This result is always more conservative than the slope condition, Eq. (8.22). For the parameters introduced in the preceding section, Eqs. (8.51) and (8.53) give a value of the stability radius about the low-temperature steady state as

$$\mathcal{A} = [(x_1 - 0.998)^2 + (x_2 - 1.001)^2]^{1/2} < 0.162$$

This compares quite favorably with the value of 0.197 obtained directly from the quadratic Liapunov function. We have already seen that the quadratic Liapunov function cannot be applied to the high-temperature steady state for the parameters used in this example.

8.7 CONCLUDING REMARKS

The intent of this chapter has been to illustrate the approaches available for the analysis of process stability. While we have considered only lumped systems, described by ordinary differential equations, the same principles apply to processes described by partial differential equations, though with somewhat more mathematical complexity. The Liapunov stability approach is quite elegant, but it is very limited in application, and the computation of the eigenvalues of the linearized process equations remains the most frequently employed method. It is therefore relatively straightforward (in principle) to establish the conditions for stability of a process model to infinitesimal perturbations, but it is rarely possible to obtain information about finite disturbances.

BIBLIOGRAPHICAL NOTES

This chapter closely follows sections of

> M. M. Denn, *Stability of Reaction and Transport Processes* (Englewood Cliffs, NJ: Prentice-Hall, Inc., 1975).

The general theory is dealt with in many other texts, including

> W. Hahn, *Stability of Motion* (New York, NY: Springer Verlag, 1967).
>
> J. LaSalle and S. Lefschetz, *Stability by Liapunov's Direct Method* (New York, NY: Academic Press, 1961).

The basic source is Liapunov's 1898 memoir, most readily accessible in the reprint of the French translation:

> A. A. Liapunov, Problème Général de la Stabilité du Mouvement, *Ann. Math. Study No. 17* (Princeton, NJ: Princeton University Press, 1949).

Applications to chemical reactors are covered extensively in

> D. D. Perlmutter, *Stability of Chemical Reactors* (Englewood Cliffs, NJ: Prentice-Hall, 1972).

9

MODAL ANALYSIS

9.1 INTRODUCTION

Process models will usually consist of a large number of nonlinear equations. There are no general methods for dealing with nonlinear systems, and each situation must be considered individually. If we are willing to restrict attention to a small region about a given operating point, however, then it is often possible to approximate the nonlinear system by an equivalent linear one; we have already utilized this approximation in the preceding chapter in order to study stability to infinitesimal disturbances.

The theory of linear systems is highly developed, and a variety of very general techniques exist for studying the qualitative behavior as well as detailed quantitative model response. All are essentially equivalent, and depend on a knowledge of the eigenvalues and eigenvectors (or eigenfunctions) of the linear system. We shall describe an approach known as *modal analysis,* which we find to be a particularly useful way of analyzing dynamical systems. We shall introduce modal analysis for a restricted class of lumped-parameter systems, but it has general applicability, including distributed-parameter processes. We shall also show how some limited information can be obtained by analyzing the dynamical modes of slightly nonlinear processes.

9.2 LINEARIZATION

Consider the dynamical system described by the ordinary differential equations

9.2 LINEARIZATION

$$\dot{\mathbf{x}} = \mathbf{f}(\mathbf{x}, \mathbf{U}) \tag{9.1}$$

$\mathbf{x} = (x_1, x_2, \ldots, x_N)$ is the set of characterizing variables, while $\mathbf{U} = (U_1, U_2, \ldots, U_m)$ is a set of possibly time-dependent forcing functions. The $\{U_i\}$ might be operator-selected control variables—flow rates to a reactor, for example.

For a given set of constant values of the $\{U_i\}$, denoted $\mathbf{U}_s = (U_{1s}, U_{2s}, \ldots, U_{Ms})$, there is a corresponding steady-state value of $\mathbf{x}$ which is denoted $\mathbf{x}_s = (x_{1s}, x_{2s}, \ldots, x_{Ns})$. $\mathbf{x}_s$ is a solution of the algebraic equations

$$\mathbf{f}(\mathbf{x}_s, \mathbf{U}_s) = 0 \tag{9.2}$$

We now consider situations in which $\mathbf{x}$ and $\mathbf{U}$ are always close to $\mathbf{x}_s$ and $\mathbf{U}_s$ in a sense that is not yet defined. We define time-dependent perturbation variables as follows:

$$\boldsymbol{\xi}(t) = \mathbf{x}(t) - \mathbf{x}_s; \quad \xi_i(t) = x_i(t) - x_{is}, \quad i = 1, 2, \ldots, N \tag{9.3a}$$

$$\mathbf{u}(t) = \mathbf{U}(t) - \mathbf{U}_s; \quad u_j(t) = U_j(t) - U_{js}, \quad j = 1, 2, \ldots, M \tag{9.3b}$$

If $\mathbf{f}$ is twice differentiable with respect to both $\mathbf{x}$ and $\mathbf{V}$, then we may expand Eq. (9.1) in a Taylor series about $\mathbf{x} = \mathbf{x}_s$, $\mathbf{U} = \mathbf{U}_s$ to obtain

$$\dot{\boldsymbol{\xi}} = \mathbf{A} \cdot \boldsymbol{\xi} + \mathbf{B} \cdot \mathbf{u} + o(\boldsymbol{\xi}, \mathbf{u}) \tag{9.4a}$$

$$\dot{\xi}_i = \sum_{j=1}^{n} A_{ij} \xi_j + \sum_{k=1}^{m} B_{ik} u_k + o(\boldsymbol{\xi}, \mathbf{u}) \tag{9.4b}$$

The symbol $o(\boldsymbol{\xi}, \mathbf{u})$ denotes terms that approach zero faster than $\boldsymbol{\xi}$ or $\mathbf{u}$ as $\boldsymbol{\xi}$ and $\mathbf{u}$ both tend to zero. The elements of the matrices $\mathbf{A}$ and $\mathbf{B}$ are defined by

$$A_{ij} = \frac{\partial f_i(\mathbf{x}_s, \mathbf{U}_s)}{\partial x_j}; \quad B_{ik} = \frac{\partial f_i(\mathbf{x}_s, \mathbf{U}_s)}{\partial U_k} \tag{9.5}$$

As in the preceding chapter, we are using the "dot product" notation of vector analysis for the inner product in place of the more common matrix notation. This permits us to neglect the distinction between column and row vectors.*

---

*$\mathbf{A} \cdot \boldsymbol{\xi}$ is a vector whose ith component is $\sum_{j=1}^{N} A_{ij} \xi_j$. $\boldsymbol{\xi} \cdot \mathbf{A}$ is a vector whose jth component is $\sum_{i=1}^{N} \xi_i A_{ij}$.

For sufficiently small values of $|\xi|$ and $|u|$ we may neglect the $o(\xi,u)$ terms in Eq. (9.4) and examine the linear approximation

$$\dot{\xi} = A \cdot \xi + B \cdot u \qquad (9.6)$$

As we shall see subsequently, this approximation has a further requirement that no eigenvalue of A have a zero real part, or else there will be some cases in which the transient of the full nonlinear system is determined by the higher-order terms. We assume for simplicity in the presentation here that all eigenvalues of A are real and distinct.

9.3 EIGENVALUES AND EIGENVECTORS

Consider the homogeneous linear differential equation

$$\dot{\xi} = A \cdot \xi \qquad (9.7a)$$

$$\dot{\xi}_i = \sum_{j=1}^{N} A_{ij} \xi_j \qquad (9.7b)$$

This equation has a solution

$$\xi(t) = \sum_{n=1}^{N} C_n e^{\lambda_n t} y_n \qquad (9.8)$$

The $\{C_n\}$ are determined from the initial conditions; $\{y_n\}$ are N constant vectors, known as *eigenvectors*, and $\{\lambda_n\}$ are the N *eigenvalues*. Substitution of Eq. (9.8) into (9.7) gives the following sequence of steps:

$$\dot{\xi} = \sum_{n=1}^{N} \lambda_n C_n e^{\lambda_n t} y_n = A \cdot \xi = \sum_{n=1}^{N} C_n e^{\lambda_n t} A \cdot y_n \qquad (9.9)$$

For this equality to hold for all t and for all initial conditions (i.e., for all $\{C_n\}$), it then follows that λ_n and y_n must satisfy the following linear algebraic system:

$$A \cdot y_n = \lambda_n y_n, \quad n = 1, 2, \ldots, N \qquad (9.10a)$$

or, equivalently,

$$(A - \lambda_n I) \cdot y_n = 0, \quad n = 1, 2, \ldots, N \qquad (9.10b)$$

9.3 EIGENVALUES AND EIGENVECTORS

Here $\mathbf{I}$ is the identity matrix, with ones on the diagonal and zeros elsewhere.

Equation (9.10) is a linear, homogeneous algebraic equation. A solution exists if and only if the determinant of the matrix $\mathbf{A} - \lambda_n \mathbf{I}$ vanishes:

$$\det (\mathbf{A} - \lambda_n \mathbf{I}) = 0 \qquad (9.11)$$

This is an Nth-order polynomial equation that determines the N eigenvalues $\{\lambda_n\}$; as noted previously, we take the eigenvalues here to be real and distinct.

For each eigenvalue λ_n, $n = 1, 2, \ldots, N$, there is an eigenvector $\mathbf{y}_n$ determined by Eq. (9.10). One component of this vector can be specified arbitrarily. It is most convenient to take the eigenvalues to be normalized to unity:

$$\mathbf{y}_n \cdot \mathbf{y}_n = \sum_{i=1}^{N} y_{ni} y_{ni} = 1 \qquad (9.12)$$

As long as the eigenvalues are distinct, the eigenvectors are linearly independent; i.e., no eigenvector can be formed from a linear combination of the others.

The *adjoint* differential equation system is also of significance:

$$\dot{\boldsymbol{\eta}} = \boldsymbol{\eta} \cdot \mathbf{A} \qquad (9.13a)$$

$$\dot{\eta}_i = \sum_{j=1}^{N} \eta_j A_{ji} \qquad (9.13b)$$

The solution to this equation is

$$\boldsymbol{\eta}(t) = \sum_{n=1}^{N} C_n^A e^{\lambda_n t} \mathbf{y}_n^A \qquad (9.14)$$

The adjoint eigenvectors satisfy the equations

$$\mathbf{y}_n^A \cdot \mathbf{A} = \lambda_n \mathbf{y}_n^A \qquad (9.15)$$

It is readily established that the eigenvalues $\{\lambda_n\}$ defined by Eq. (9.15) are identical to those defined by Eq. (9.10); this is because $\det(\mathbf{A}^\dagger - \lambda \mathbf{I})$ is identical to $\det(\mathbf{A} - \lambda \mathbf{I})$, where $\mathbf{A}^\dagger$ is the transpose of $\mathbf{A}$ (the matrix obtained by replacing A_{ij} with A_{ji}). The adjoint eigenvectors $\{\mathbf{y}_n^A\}$ are distinct from the eigenvectors $\{\mathbf{y}_n\}$, but are orthogonal to the latter*:

$$\mathbf{y}_n \cdot \mathbf{y}_m^A = \sum_{i=1}^{N} y_{ni} y_{mi}^A = 0, \quad n \neq m \tag{9.16}$$

The most convenient normalization of the adjoint set is as follows:

$$\mathbf{y}_n \cdot \mathbf{y}_n^A = \sum_{i=1}^{N} y_{ni} y_{ni}^A = 1 \tag{9.17}$$

Since the eigenvectors are linearly independent, we can express any vector of the same dimension as a linear combination of them. We do so with the solution $\boldsymbol{\xi}(t)$ of Eq. (9.7):

$$\boldsymbol{\xi}(t) = \sum_{n=1}^{N} \mathcal{A}_n(t) \mathbf{y}_n \tag{9.18}$$

Then

$$\dot{\boldsymbol{\xi}}(t) = \sum_{n=1}^{N} \dot{\mathcal{A}}_n(t) \mathbf{y}_n = \sum_{n=1}^{N} \mathcal{A}_n(t) \mathbf{A} \cdot \mathbf{y}_n$$

$$= \sum_{n=1}^{N} \lambda_n \mathcal{A}_n(t) \mathbf{y}_n \tag{9.19}$$

where we have used Eq. (9.10) in the last substitution. It then follows that the functions $\mathcal{A}_n(t)$ satisfy the equations

$$\dot{\mathcal{A}}_n = \lambda_n \mathcal{A}_n, \quad n = 1, 2, \ldots, N \tag{9.20}$$

That is, *the dynamical system has been uncoupled into N first-order equa-*

---

* $\mathbf{A} \cdot \mathbf{y}_n = \lambda_n \mathbf{y}_n, \quad n = 1, 2, \ldots, N$

$\mathbf{y}_m^A \cdot \mathbf{A} = \lambda_m \mathbf{y}_m^A, \quad m = 1, 2, \ldots, N$

Premultiplying the first equation by $\mathbf{y}_m^A$ and postmultiplying the second by $\mathbf{y}_n$ and then subtracting, we obtain

$$0 = (\lambda_n - \lambda_m) \mathbf{y}_n \cdot \mathbf{y}_m^A$$

Since $\lambda_n \neq \lambda_m$, $n \neq m$, Eq. (9.16) follows immediately.

tions representing the responses of individual modes. The inverse relation between the characterizing variables $\boldsymbol{\xi}(t)$ and the modal variables $\{\mathcal{A}_n(t)\}$ is established by multiplying Eq. (9.18) by each adjoint eigenvector and making use of the orthogonality, Eq. (9.16):

$$\mathcal{A}_m(t) = \mathbf{y}_m^A \cdot \boldsymbol{\xi}(t) = \sum_{i=1}^{N} y_{mi}^A \xi_i(t) \tag{9.21}$$

Equation (9.21) provides the initial values $\{\mathcal{A}_n(0)\}$ corresponding to initial conditions $\boldsymbol{\xi}(0)$.

9.4 NONHOMOGENEOUS SYSTEMS

We can now return to the general nonhomogeneous linear system defined by Eq. (9.6). The transformation $\boldsymbol{\xi} = \Sigma \mathcal{A}_n \mathbf{y}_n$, with $\{\mathbf{y}_n\}$ the eigenvectors of $\mathbf{A}$, then leads to

$$\sum_{n=1}^{N} \dot{\mathcal{A}}_n \mathbf{y}_n = \sum_{n=1}^{N} \mathcal{A}_n \mathbf{A} \cdot \mathbf{y}_n + \mathbf{B} \cdot \mathbf{u}$$

$$= \sum_{n=1}^{N} \mathcal{A}_n \lambda_n \mathbf{y}_n + \mathbf{B} \cdot \mathbf{u} \tag{9.22}$$

Scalar multiplication with $\mathbf{y}_m^A$ then yields

$$\dot{\mathcal{A}}_m = \lambda_m \mathcal{A}_m + (\mathbf{y}_m^A \cdot \mathbf{B}) \cdot \mathbf{u}, \quad m = 1, 2, \ldots, N \tag{9.23}$$

That is, the dynamical modes are uncoupled in the same manner as for the homogeneous system, but the nonhomogeneous term remains in each dynamical mode.

9.5 ORDER REDUCTION

Modal analysis provides an effective means of reducing the order of a dynamical system. Suppose that all eigenvalues of $\mathbf{A}$ from $n = \nu + 1$ to $n = N$ are large negative numbers, and the dynamical response is therefore dominated by the first ν modes. We can construct a dynamical system of order ν having the identical responses for the first ν modes as the full Nth-order dynamical system by noting that $\mathcal{A}_n$ responds very quickly for $\nu + 1 \leq n$

≤ N, and appears always to be at steady state on a time scale defined by the first v dynamical modes. Thus, $\dot{\mathcal{A}}_n \approx 0$, $v + 1 \leq n \leq N$, and we can write the final $N - v$ components of Eq. (9.23) as

$$\mathcal{A}_n = -\frac{1}{\lambda_n}(\mathbf{y}_n^A \cdot \mathbf{B}) \cdot \mathbf{u}, \quad v + 1 \leq n \leq N \tag{9.24a}$$

The remaining modes are described by the uncoupled first-order equations

$$\dot{\mathcal{A}}_n = \lambda_n \mathcal{A}_n + (\mathbf{y}_n^A \cdot \mathbf{B}) \cdot \mathbf{u}, \quad 1 \leq n \leq v \tag{9.24b}$$

The variables $\boldsymbol{\xi}(t)$ are then recovered by solving Eqs. (9.24b) and using Eq. (9.21):

$$\boldsymbol{\xi}(t) = \sum_{n=1}^{v} \mathcal{A}_n(t)\mathbf{y}_n - \sum_{n=v+1}^{N} \frac{1}{\lambda_n} \mathbf{y}_n (\mathbf{y}_n^A \cdot \mathbf{B}) \cdot \mathbf{u} \tag{9.25}$$

This approach to order reduction based on the magnitudes of the eigenvalues is powerful and very straightforward, and it is often used to reduce the complexity of high-order models. It has one major deficiency, in that the transient response of a nonhomogeneous system is governed by the forcing terms as well as the eigenvalues. As an extreme example, suppose that all components of the vectors $\mathbf{y}_m^A \cdot \mathbf{B}$ were to be zero for $1 \leq m \leq v$. In that case, the forcing terms would have no effect at all on the responses of the slowest (first v) modes, and these modes would die out. The entire response would then be determined by modes $v + 1$ to N, and the pseudo-steady state response ($\dot{\mathcal{A}}_m \approx 0$) for these modes would be inappropriate. The coefficients of the forcing terms must therefore be taken into account in selecting the v dynamical modes to be retained, and the decision cannot be made solely on the basis of the eigenvalues. (In control-theory terminology, the zeros as well as the poles of the transfer function matrix must be considered. Dynamical modes for which $\mathbf{y}_m^A \cdot \mathbf{B}$ is identically zero are said to be *uncontrollable*, since they cannot be affected by any external forcing.)

9.6 MODAL CONTROL

Modal analysis has been used extensively for the design of control systems to regulate multivariable processes. We shall illustrate the simplest procedure here; there are subtleties that arise in applications to real processes, but the basic principles do not change.

We shall suppose that the components of $\mathbf{u}(t)$ in Eq. (9.6) are all variables which we may specify for control purposes. If $\mathbf{u}$ has M components, then we may control the response of M dynamical modes independently. Let us suppose that these are the first M modes, and that we wish the mth mode to respond like a first-order system with eigenvalue Λ_m; the first M components of Eq. (9.23) are then

$$\dot{\mathcal{A}}_m = \lambda_m \mathcal{A}_m + (\mathbf{y}_m^A \cdot \mathbf{B}) \cdot \mathbf{u} = \Lambda_m \mathcal{A}_m, \quad m = 1, 2, \ldots, M \quad (9.26)$$

or

$$(\mathbf{y}_m^A \cdot \mathbf{B}) \cdot \mathbf{u} = (\Lambda_m - \lambda_m) \mathcal{A}_m$$
$$= (\Lambda_m - \lambda_m) \mathbf{y}_m^A \cdot \boldsymbol{\xi}(t), \quad m = 1, 2, \ldots, M \quad (9.27)$$

This provides M equations with which to solve for the M components of $\mathbf{u}$ in terms of the N characterizing variables $\boldsymbol{\xi}$. We are thus led to a linear multivariable feedback control system. The selection of the closed-loop eigenvalues $\{\Lambda_m\}$ is arbitrary, but it is obvious that they should not differ significantly from one another, nor should they be more negative than the eigenvalue associated with any uncontrolled mode.

9.7 FLUID CATALYTIC CRACKER

9.7.1 Linear Model

Kurihara's model of the FCC was introduced in Section 5.9. The five nonlinear ordinary differential equations can be linearized about the steady state to give a linear system in the form of Eq. (9.6), with the following definitions of the components of $\boldsymbol{\xi}$ and $\mathbf{u}$:

$$\boldsymbol{\xi} = \begin{bmatrix} \delta R_{ra} \\ \delta T_{rg} \\ \delta C_{rc} \\ \delta C_{sc} \\ \delta C_{cat} \end{bmatrix}, \quad \mathbf{u} = \begin{bmatrix} \delta R_{ai} \\ \delta R_{rc} \end{bmatrix} \quad (9.28)$$

The symbol "δ" denotes the difference from the steady-state value given in Table 5.1. The matrices $\mathbf{A}$ and $\mathbf{B}$ have the following values for the parameters in Table 5.1:

$$\mathbf{A} = \begin{bmatrix} -73.3 & 40.18 & 341.0 & 0.0 & 1551.8 \\ 11.97 & -12.0 & 142.6 & 0.0 & 0.0 \\ 0.0 & -0.0029 & -12.3 & -12.0 & 0.0 \\ 0.163 & 0.0 & 36.63 & -40.0 & -40.12 \\ 0.163 & 0.0 & -3.5 & 0.0 & -80.0 \end{bmatrix} \quad (9.29a)$$

$$\mathbf{B} = \begin{bmatrix} 0.0 & 230.6 \\ 6.90 & -68.72 \\ -0.028 & 0.28 \\ 0.0 & -0.878 \\ 0.0 & -0.875 \end{bmatrix} \quad (9.29b)$$

The deviation in flue gas oxygen is given by linearization of Eq. (5.62b):

$$\delta O_{fg} = -5.77 \times 10^{-3} \delta T_{rg} - 0.797 \delta C_{rc} + 4.57 \times 10^{-5} \delta R_{ai} \quad (9.30)$$

The eigenvalues of the matrix $\mathbf{A}$ are $(-0.198, -5.014, -52.46, -64.99, -94.93)$. The presence of three negative eigenvalues with very large magnitudes indicates that there is an equivalent second-order system that generally preserves the important transient response of the full linear model. This observation is consistent with Kurihara's approximation that was described in Section 5.9 and illustrated in Fig. 5.12 for one particular transient. The second-order system could be constructed systematically using the procedure in Section 9.5 in order to have eigenvalues identical to the first two eigenvalues of the full fifth-order process model. In this particular case our understanding of the process has enabled us to identify two physical variables whose dynamics dominate the response; it is therefore convenient to follow Kurihara, as in Section 5.9.5, and suppose that the response of the reactor is sufficiently rapid to enable us to set $d\delta T_{ra}/dt$, $d\delta C_{sc}/dt$, and $d\delta C_{cat}/dt$ to zero.* This leads to three linear algebraic equations relating the five elements of $\boldsymbol{\xi}$ and the two of $\mathbf{u}$; we record only one here for subsequent use:

$$\delta T_{ra} = 0.608 \delta T_{rg} - 4.848 \delta O_{fg} + 3.058 \delta R_{rc} + 2.21 \times 10^{-4} \delta R_{ai} \quad (9.31)$$

Substitution of these linear algebraic relations into Eqs. (9.29) leads to a second-order model entirely in terms of regenerator variables:

$$\dot{\hat{\boldsymbol{\xi}}} = \hat{\mathbf{A}} \cdot \hat{\boldsymbol{\xi}} + \hat{\mathbf{B}} \cdot \mathbf{u} \quad (9.32)$$

$$\hat{\boldsymbol{\xi}} = \begin{pmatrix} \delta T_{rg} \\ \delta C_{rc} \end{pmatrix} \quad (9.33)$$

$$\hat{\mathbf{A}} = \begin{pmatrix} -5.14 & 189 \\ 1.13 \times 10^{-2} & -0.692 \end{pmatrix} \quad (9.34a)$$

---

*This is equivalent to having $\mathcal{A}_1$ and $\mathcal{A}_2$ consist of a linear combination of only δT_{rg} (ξ_2) and δC_{rc} (ξ_3). According to Eq. (9.21), $\mathbf{y}_1^A$ and $\mathbf{y}_2^A$ must therefore be dominated by the second and third components, which is in fact true.

9.7 FLUID CATALYTIC CRACKER

$$\hat{B} = \begin{pmatrix} 6.90 & -32.3 \\ -2.63 \times 10^{-2} & 0.206 \end{pmatrix} \quad \text{(9.34b)}$$

The eigenvalues of $\hat{A}$ are $(-0.261, -5.575)$, indicating that the dominant dynamics of the reduced-order model are close to those of the fifth-order system. The adjoint eigenvectors are

$$\mathbf{y}_1^A = \begin{pmatrix} 0.079 \\ 35.71 \end{pmatrix}, \quad \mathbf{y}_2^A = \begin{pmatrix} -0.922 \\ 35.71 \end{pmatrix} \quad \text{(9.35)}$$

9.7.2 Control

Closed-loop eigenvalues are a major factor in determining the transient response of the controlled system, and they provide a useful means of screening control algorithms. We consider here only linear proportional controllers, in which the control changes δR_{ai} and δR_{rc} are linear in the measured deviations of the system variables. Table 9.1 lists five policies that have been proposed for the Kurihara model of the FCC. The "conventional" control is as given by Kurihara and shown in Eqs. (5.63); for purposes of comparison we do not include the integral (*reset*) mode that would normally be a part of such a control system. Kurihara's suboptimal control, given as Eqs. (5.64), was illustrated in Fig. 5.14. The controls synthesized by Nakano are all ratio controls ($\delta R_{ai}/\delta R_{rc}$ = constant) as a consequence of the algorithm that he used. It is clear that the second-order model is also reasonably successful in representing the closed-loop behavior of the fifth-order system. With the exception of the conventional control, all of the quite different control strategies lead to a dynamic response that is at least a factor of twenty faster than the uncontrolled system, with dominant eigenvalues having real parts of order -6.

The modal variables for Eqs. (9.32) through (9.35) are obtained from Eq. (9.21) as

$$\mathcal{A}_1 = 0.079 \delta T_{rg} + 35.71 \delta C_{rc} \quad \text{(9.36a)}$$

$$\mathcal{A}_2 = -0.922 \delta T_{rg} + 35.71 \delta C_{rc} \quad \text{(9.36b)}$$

and the uncoupled system equations (9.23) are

$$\dot{\mathcal{A}}_1 = -0.261 \mathcal{A}_1 + (-0.394 \delta R_{ai} + 4.80 \delta R_{rc}) \quad \text{(9.37a)}$$

$$\dot{\mathcal{A}}_2 = -5.58 \mathcal{A}_2 + (-7.30 \delta R_{ai} + 37.08 \delta R_{rc}) \quad \text{(9.37b)}$$

TABLE 9.1.
Closed-loop eigenvalues for FCC

Control	Feedback Law	Eigenvalues			
		2nd-Order Model		5th-Order Model	
		Real	Imag.	Real	Imag.
Open loop	$\delta R_{ai} = 0$	− 0.261	0	− 0.198	0
	$\delta R_{rc} = 0$	− 5.575	0	− 5.014	0
				− 52.463	0
				− 64.992	0
				− 94.932	0
"Conventional"	$\delta R_{ai} = -40\delta O_{fg}$	− 1.443	−1.998	− 1.266	− 1.799
	$\delta R_{rc} = -0.28\delta T_{ra}$	− 1.443	1.998	− 1.266	1.799
				− 52.741	0
				− 71.208	0
				−136.548	0
Kurihara	$\delta R_{ai} = -1.0\ \delta T_{rg}$	− 5.776	−4.771	− 5.426	− 4.501
	$\delta R_{rc} = +50.0\delta O_{fg}$	− 5.776	4.771	− 5.426	4.501
				− 53.508	− 8.091
				− 53.508	8.091
				− 97.806	0

Nakano No. 1	$\delta R_{ai} = 0$	-7.055	0	-6.587	0
		-11.850	0	-10.813	0
	$\delta R_{rc} = 0.5018 T_{rg} + 139 \delta O_{fg}$			-49.912	-14.301
				-49.912	-14.301
				-110.707	0
Nakano No. 6	$\delta R_{ai} = 1.59 \delta T_{rg} + 537 \delta O_{fg}$	-8.519	-1.537	-6.570	0
		-8.519	1.537	-11.150	0
	$\delta R_{rc} = 0.476 \delta T_{rg} + 161 \delta O_{fg}$			-46.694	-12.630
				-46.694	12.630
				-109.802	0
Nakano No. 7	$\delta R_{ai} = 13.4 \delta T_{rg} + 1960 \delta O_{fg}$	-6.106	0	-8.167	0
		-24.446	0	-10.267	0
	$\delta R_{rc} = 2.69 \delta T_{rg} + 391 \delta O_{fg}$			-50.037	-14.536
				-50.037	14.536
				-163.681	0

The control equations (9.27) in terms of the desired eigenvalues Λ_1 and Λ_2 of the controlled system are then

$$-0.394\delta R_{ai} + 4.80\delta R_{rc} = (\Lambda_1 + 0.261)(0.079\delta T_{rg} + 35.71\delta C_{rc}) \quad \textbf{(9.38a)}$$

$$-7.30\delta R_{ai} + 37.08\delta R_{rc} = (\Lambda_2 + 5.58)(-0.922\delta T_{rg} + 35.71\delta C_{rc}) \quad \textbf{(9.38b)}$$

For given Λ_1 and Λ_2, these two equations give δR_{ai} and δR_{rc} as linear combinations of δT_{rg} and δC_{rc}.

Suppose, for example, that the response of the second mode without control is considered to be sufficiently rapid, in which case we may set $\Lambda_2 = -5.58$. Equation (9.38b) then establishes a fixed ratio between the two flow rate controllers:

$$\Lambda_2 = -5.58: \quad \delta R_{ai} = 5.08\delta R_{rc} \quad \textbf{(9.39a)}$$

and Eq. (9.39a) becomes

$$\Lambda_2 = -5.58: \quad \delta R_{rc} = -(\Lambda_1 + 0.26)(0.028\delta T_{rg} + 12.75\delta C_{rc}) \quad \textbf{(9.39b)}$$

If we wish both modes to respond at the same rate, then we should also set Λ_1 to -5.58 to obtain

$$\Lambda_1 = \Lambda_2 = -5.58: \quad \delta R_{rc} = 0.15\delta T_{rg} + 68\delta C_{rg} \quad \textbf{(9.39c)}$$

δO_{fg} is much more likely to be measured than δC_{rc}, so this control scheme is rewritten with the use of Eq. (9.30) to give

$$M1: \quad \delta R_{ai} = 1.73\delta T_{rg} + 433\delta O_{fg}$$

$$\delta R_{rc} = 0.34\delta T_{rg} + 85\delta O_{fg} \quad \textbf{(9.40)}$$

The eigenvalues of the closed-loop system are shown in Table 9.2. The response of the full fifth-order process model with this control system is just slightly slower than predicted with the reduced second-order model.

A faster response can be obtained by increasing the magnitudes of Λ_1 and Λ_2. If, for example, we set $\Lambda_1 = \Lambda_2 = -8$, then the flow changes are no longer in a constant ratio and we obtain from Eqs. (9.30) and (9.38)

$$M2: \quad \delta R_{ai} = 1.85\delta T_{rg} + 605\delta O_{fg}$$

$$\delta R_{rc} = 0.45\delta T_{rg} + 122\delta O_{fg} \quad \textbf{(9.41)}$$

The computed closed-loop eigenvalues are shown in Table 9.2. The differ-

TABLE 9.2.
Closed-loop eigenvalues for FCC.

Control	Feedback Law	2nd-Order Model Real	2nd-Order Model Imag.	5th-Order Model Real	5th-Order Model Imag.
M1	$\delta R_{ai} = 1.73\delta T_{rg} + 433\delta O_{fg}$	−5.563	0	−4.871	0
		−5.617	0	−5.229	0
	$\delta R_{rc} = 0.34\delta T_{rg} + 85\delta O_{fg}$			−52.911	−8.704
				−52.911	8.704
				−105.960	0
M2	$\delta R_{ai} = 1.85\delta T_{rg} + 605\delta O_{fg}$	−8.211	−0.07	−7.297	0
		−8.211	0.07	−7.983	0
	$\delta R_{rc} = 0.48\delta T_{rg} + 122\delta O_{fg}$			−50.440	−10.062
				−50.440	10.062
				−108.909	0

ence from -8.0 for the closed-loop reduced-order system reflects the rounding errors because of the unequal magnitudes of matrix elements.

9.7.3 Measurement and Response

The model reduction expresses the dynamical response entirely in terms of the regenerator dynamics. Yet, control of the reactor is a primary concern of the operator, and it appears that the reactor is left uncontrolled with schemes like $M1$ and $M2$. Thus, such a control strategy might fail the critical test of operator acceptability. The apparent neglect of the reactor is, of course, illusory, as Eq. (9.31) shows, since the reactor and regenerator are tightly coupled. Thus, effective control of the reactor will follow from effective regenerator control. The modal control laws can be expressed in terms of reactor temperature and flue gas oxygen as

$$M1': \quad \delta R_{ai} = 1.10 \delta T_{ra} + 142 \delta O_{fg}$$
$$\delta R_{rc} = 0.22 \delta T_{ra} + 28 \delta O_{fg} \quad (9.40')$$
$$M2': \quad \delta R_{ai} = 0.93 \delta T_{ra} + 256 \delta O_{fg}$$
$$\delta R_{rc} = 0.24 \delta T_{ra} + 32 \delta O_{fg} \quad (9.41')$$

The multivariable nature of each of the controls prevents specific identification of either control variable with regulation of the reactor temperature, however.

We could continue to analyze the properties of Eq. (9.38) to gain further insight into the controller design problem, but that is not our primary purpose here. It can be shown that a control law

$$\delta R_{rc} = K_r \delta T_{ra}$$
$$\delta R_{ai} = K_a \delta O_{fg} \quad (9.42)$$

will emerge as a close approximation to the computed multivariable system as Λ_1 increases with Λ_2 very large in magnitude, tending to $K_r = 0.33$ and $K_a = 48 - 71\Lambda_1$. The fact that δR_{ai} will be dominated by δO_{fg} is suggested by control laws $M1$ and $M2$. As shown in Figs. 5.13 and 5.14, we expect δO_{fg} to be of order 0.1 or so, while δT_{ra} will be of order 10; the oxygen term is increasing rapidly in importance relative to the reactor temperature term as the dominant eigenvalue goes from -5.6 to -8, and is already responsible for most of the change in R_{ai}. The other eigenvalues of the fifth-order system become the dominant ones for this scheme, however, when K_a reaches about 600, and the performance deteriorates with further increase in K_a. This uncoupling of the control into single loops, with one process measurement setting one control variable, has considerable intuitive appeal; it is clearly irrelevant from the point of view of an automated control system, however, even one implemented with analog devices having no logical circuits.

The controllers synthesized using modal analysis have positive feedback gains (i.e., positive coefficients on the process deviations from steady state), while the conventional controller utilizes negative feedback. The difference in mechanism leads to a much more rapid response, and is best understood by focusing on the reactor temperature.

The gas oil conversion reaction is endothermic, in that heat must be supplied to maintain the temperature. The reactor temperature can therefore be reduced either by increasing the rate at which the reaction occurs or by bringing colder solids into the reactor. (The latter will reduce the reaction rate, but by lowering the temperature through a sensible heat effect.) In the conventional scheme, the reactor temperature is reduced by the catalyst flow rate. The reduced catalyst rate increases the catalyst residence time in the reactor, thus increasing the rate of conversion and hence reducing the temperature. Simultaneously, however, the carbon deposition rate is increased and the regenerator residence time is increased; both effects will lead to increased regenerator reaction and temperature, thus bringing hotter catalyst into the reactor and tending to resist the reactor temperature control. The increased reaction in the regenerator will lower the flue gas oxygen, but the air rate will be increased to maintain this level.

The positive feedback policies, on the other hand, exploit the regenerator dynamics directly. An increased reactor or regenerator temperature calls for an *increase* in catalyst rate. The immediate effect on the reactor may be a slight temperature increase, but the reduced residence time in the regenerator results in colder catalyst coming into the reactor and an ultimate decrease in reactor temperature. The role of an increased air rate following an increase in flue gas oxygen is more subtle. The effect seems to be one of ensuring sufficient catalyst activity to maintain the reaction.

Some caution needs to be used in the application of multivariable control laws like those obtained here. Failure of one sensor (i.e., sending a zero-error signal to the controller) will cause the process to become unstable; this is a common occurrence in the use of advanced control techniques for designing multivariable controllers, and is related to the most efficient use of the system dynamics. Furthermore, it is important to reiterate that these results are only preliminary, in that they examine only the closed-loop eigenvalues (poles). System zeros are important in the propagation of disturbances, as noted in Section 9.5, and they must be accounted for in any final controller design. Finally, with regard to the specific application discussed here, we repeat the qualification stated in Section 5.9.1: We have been concerned only with the sluggish response of the FCC, and have entirely disregarded both the need to adjust operating characteristics to meet varying process requirements and the possible effect of product variations on downstream processing units. Clearly these concerns cannot be ignored in a real application.

9.8 SLIGHTLY NONLINEAR PROCESSES

Modal analysis has not often been applied to nonlinear systems, but it has the potential for substantial order reduction in certain systems of high order that are weakly nonlinear. The application most widely studied has been the nonlinear stability of fluid mechanical systems. We shall develop the basic ideas here and illustrate the approach with an example of an overly simple model of a fluidized bed reactor. For this example the order reduction afforded by a modal decomposition enables us to obtain an analytical solution of a "stiff" system of differential equations that has proved to be difficult to integrate by standard numerical methods.

9.8.1 Quadratic Expansion

Suppose that $\mathbf{f}$ in Eq. (9.1) depends only on $\mathbf{x}$, and is at least twice differentiable. If we expand in a Taylor series about the origin and retain terms to at least *second* order, then we obtain

$$\dot{\xi}_i = \sum_{j=1}^{N} A_{ij}\xi_j + \sum_{j,k=1}^{N} C_{ijk}\xi_j\xi_k + o(|\xi|^2) \tag{9.43}$$

$\mathbf{A}$ is defined by Eq. (9.5), while

$$C_{ijk} = \frac{1}{2}\frac{\partial^2 f_i(\mathbf{x}_s)}{\partial x_j\, \partial x_k} \tag{9.44}$$

Introducing the modal functions $\mathcal{A}_n(t)$ defined by Eq. (9.18), and taking the inner ("dot") product of each term in Eq. (9.43) with $\mathbf{y}_p^A$, we obtain

$$\dot{\mathcal{A}}_p = \lambda_p \mathcal{A}_p + \sum_{n,m=1}^{N} I_{pnm}\mathcal{A}_n\mathcal{A}_m + o(|\mathcal{A}|^2) \tag{9.45}$$

$$I_{pnm} = \sum_{i,j,k=1}^{N} C_{ijk} y_{pi}^A y_{nj} y_{mk} \tag{9.46}$$

9.8 SLIGHTLY NONLINEAR PROCESSES

We now make several assumptions. The first is that the eigenvalues $\{\lambda_p\}$ are real, distinct, and negative, so that they can be ordered $0 > \lambda_1 > \lambda_2 > \cdots$. This limits consideration to steady states that are stable to infinitesimal perturbations. We also assume that the eigenvalue λ_1 is much smaller in magnitude than any other eigenvalue and dominates the linear response:

$$|\lambda_1/\lambda_p| \ll 1, \quad p = 2, 3, \ldots, N \tag{9.47}$$

This is a critical assumption, as is the next: that the modal variables $\{\mathcal{A}_p(t)\}$ are roughly ordered in magnitude according to the reciprocals of the eigenvalues; i.e.,

$$|\lambda_p \mathcal{A}_p| \sim |\lambda_q \mathcal{A}_q|, \quad p, q = 1, 2, \ldots, N \tag{9.48}$$

According to Eq. (9.48) each term in the sum in Eq. (9.45) is roughly

$$I_{pnm} \frac{\lambda_1^2}{\lambda_n \lambda_m} \mathcal{A}_1^2$$

Only the term corresponding to $n = m = 1$ is significant (as long as I_{p11} is finite), according to the assumption in Eq. (9.47). Thus, Eq. (9.45) simplifies to

$$\dot{\mathcal{A}}_p = \lambda_p \mathcal{A}_p + I_{p11}\mathcal{A}_1^2, \quad p = 1, 2, \ldots, N \tag{9.49}$$

That is, the equation for $\mathcal{A}_1(t)$ contains a quadratic nonlinearity, while all other modal variables are governed by a nonhomogeneous linear equation.

Equation (9.47) has an analytical solution for each p:

$$\mathcal{A}_1(t) = \mathcal{A}_1(0) \exp(\lambda_1 t) \left\{ 1 + \frac{\mathcal{A}_1(0) I_{111}}{\lambda_1} [1 - \exp(\lambda_1 t)] \right\}^{-1} \tag{9.50}$$

$$\mathcal{A}_p(t) = \mathcal{A}_p(0) \exp(\lambda_p t)$$
$$+ I_{p11} \int_0^t \exp[\lambda_p(t - \tau)]\mathcal{A}_1^2(\tau) d\tau, \quad p = 2, 3, \ldots, N \tag{9.51}$$

The initial conditions $\{\mathcal{A}_p(0)\}$ are computed from $\boldsymbol{\xi}(0)$ through Eq. (9.21), and $\boldsymbol{\xi}(t)$ is recovered from Eqs. (9.50) and (9.51) through application of Eq. (9.18). It is to be noted that Eqs. (9.50) and (9.51) explicitly retain the multiple time scales characteristic of the original process equations.

9.8.2 Linearization and Stability Limits

Equations (9.49) through (9.51) provide an explicit estimate of the range of validity of a linearized dynamical analysis as long as we are close to the point of marginal stability ($\lambda_1 \to 0$). The linear terms dominate only as long as

$$|\mathcal{A}_1(0)| \ll |\lambda_1/I_{111}| \tag{9.52}$$

Since the first mode dominates the response, it follows from Eq. (9.18) that $\boldsymbol{\xi} \sim \mathcal{A}_1 \mathbf{y}_1$, $\mathcal{A}_1 \sim \xi_i/y_{1i}$ for all $i = 1, 2, \ldots, N$, and Eq. (9.52) is approximately

$$|\xi_i(0)| \ll |y_{1i}||\lambda_1/I_{111}| \tag{9.53}$$

This condition is conservative for establishing the range of the linear equations, since it does not allow for the nonzero initial values of other rapidly decaying modes.

Equation (9.50) also provides an estimate of the region of stability about the steady state. Since $\lambda_1 < 0$, all deviations from $\boldsymbol{\xi} = \mathbf{0}$ decay as long as $\mathcal{A}_1(0)I_{111}/\lambda_1 > -1$. y_n is undetermined as to algebraic sign with the normalization in Eq. (9.12), and hence so is $\mathbf{y}_n^A$ with the normalization in Eq. (9.17). Thus, $\mathcal{A}_1(0)$ cannot be distinguished from $-\mathcal{A}_1(0)$ in Eq. (9.21), and only the magnitude of $\mathcal{A}_1(0)$ is of significance. In that case the condition for stability is

$$|\mathcal{A}_1(0)| = \left| \sum_{i=1}^{N} y_{1i}^A \xi_i(0) \right| < |\lambda_1/I_{111}| \tag{9.54}$$

Equation (9.54) defines a stability region bounded by two hyperplanes. Terms through third order would be required in the expansion of Eq. (9.1) to obtain an estimate of a closed stability region. The expression is generally carried out to third order for problems in fluid mechanics.

9.8.3 Batch Fluidized Bed

The requirement that the first two eigenvalues of the linearized system equations be widely separated is restrictive, but such processes do exist. The Kurihara model of the FCC is one such example, as is the binary distillation column discussed in Section 11.3. A good example for illustration is provided by a model of a batch catalytic bed published by Luss and Amundson in 1968. The physical assumptions are given in the original paper and are

9.8 SLIGHTLY NONLINEAR PROCESSES

not repeated here, though it is clear that the structure of the equations is similar to that encountered previously for well-stirred reactors. The system equations are

$$\frac{dP}{dt} = P_e - P + H_g(P_p - P) \tag{9.55a}$$

$$\frac{dT}{dt} = T_e - T + H_w(T_w - T) + H_T(T_p - T) \tag{9.55b}$$

$$A\frac{dP_p}{dt} = H_g(P - P_p) - H_p K k P_p \tag{9.55c}$$

$$C\frac{dT_p}{dt} = H_T(T - T_p) + H_T F K k P_p \tag{9.55d}$$

$$k = k_0 \exp(-\Delta E/RT_p) \tag{9.55e}$$

Here P and T denote the gas partial pressure and temperature, respectively; P_p and T_p the partial pressure and temperature in the catalyst particle; and P_e and T_e the partial pressure and temperature of the entering stream. The values of the dimensionless parameters used by Luss and Amundson are given in Table 9.3, and the corresponding steady states are tabulated in Table 9.4.

Perturbation variables are defined as follows:

$$\xi_1 = P - P_s, \qquad \xi_2 = T - T_s$$
$$\xi_3 = P_p - P_{ps}, \qquad \xi_4 = T_p - T_{ps}$$

A Taylor series expansion of Equations (9.55) through second order then gives

$$\frac{d\xi_1}{dt} = -(H_g + 1)\xi_1 + H_g\xi_3 \tag{9.56a}$$

TABLE 9.3.
Dimensionless parameters for fluidized bed.

$A = 0.17143$	$C = 205.7143$
$F = 8000$	$H_g = 320$
$H_T = 266.667$	$H_w = 1.6$
$K = 0.0024$	$Kk = 0.0006 \exp(20.7 - 15000/T_p)$

TABLE 9.4.
Steady-state solution to Eqs. (9.55) and corresponding eigenvalues of Matrix **A**.

	Steady State		
	1	2	3
P	0.09353	0.06705	0.00682
P_p	0.09351	0.06694	0.00653
T	690.445	758.346	912.764
T_p	690.607	759.169	915.093
λ_1	− 0.006315	+ 0.005695	− 0.008963
λ_2	− 0.91165	− 1.2928	− 12.563
λ_3	− 270.55	− 270.56	− 270.56
λ_4	−2187.2	−2189.5	−2258.0

$$\frac{d\xi_2}{dt} = -(1 + H_w + H_T)\xi_2 + H_T\xi_4 \tag{9.56b}$$

$$\frac{d\xi_3}{dt} = \frac{H_g}{A}\xi_1 - \left(\frac{H_pP_s}{AP_{ps}}\right)\xi_3 - \left[\frac{H_g(T_s - T_{ps})\Delta E}{AFRT_{ps}^2}\right]\xi_4$$
$$- \left[\frac{H_g(T_s - T_{ps})\Delta E(\Delta E - 2RT_{ps})}{2AFR^2T_{ps}^4}\right]\xi_4^2 \tag{9.56c}$$

$$\frac{d\xi_4}{dt} = \frac{H_T}{C}\xi_2 + \left[\frac{H_TF(P_s - P_{ps})}{CP_{ps}}\right]\xi_3$$
$$+ \frac{H_T}{C}\left[\frac{\Delta E(T_{ps} - T_s)}{RT_{ps}^2} - 1\right]\xi_4$$
$$+ \left[\frac{H_T(T_{ps} - T_s)\Delta E}{CP_{ps}RT_{ps}^2}\right]\xi_3\xi_4 \tag{9.56d}$$
$$+ \left[\frac{HT(T_{ps} - T_s)\Delta E(\Delta E - 2RT_{ps})}{2CR^2T_{ps}^4}\right]\xi_4^2$$

The eigenvalues of the matrix **A** are given in Table 9.4. It is to be noted that the first and third steady states are stable, while the second is unstable. Each of the stable states satisfies the conditions on the eigenvalues, $|\lambda_1/\lambda_p| \ll 1, p \geq 2$. The basic and adjoint eigenvectors of **A** for each of the stable steady states are recorded in Table 9.5. The only nonzero elements of **C** in Eqs. 9.43 are C_{344}, C_{434} and C_{444}.

9.8 SLIGHTLY NONLINEAR PROCESSES

Luss and Amundson obtained numerical solutions of the full nonlinear equations for a number of initial conditions. They encountered numerical difficulties because of the extremely small step size required for numerical stability, and they ultimately adopted a procedure of setting P_p and P equal when they approached to within 1% of each other. The same cases are shown here using eigenvalues and eigenvectors of the **A**-matrix and the analytical solutions embodied in Eqs. (9.18), (9.50), and (9.51). Figure 9.1 shows results for a case in which the particles are initially at the temperature of the third steady state but in which there are large deviations in gas temperatures and in both partial pressures. This case corresponds to Fig. 3 of Luss and Amundson, and the dashed line is their reported numerical solution for P_p. The results are almost identical, and the same is true for the other curves

TABLE 9.5.
Basic and adjoint eigenvectors for the two stable steady states.

MODE (j)
Steady State 1

	1	2	3	4
y_{j1}	1.3631×10^{-4}	2.6908×10^{-1}	3.5404×10^{-9}	-1.690×10^{-1}
y_{j2}	-7.0368×10^{-1}	-6.5183×10^{-1}	-9.999×10^{-1}	-1.385×10^{-4}
y_{j3}	1.3673×10^{-4}	2.6915×10^{-1}	5.5816×10^{-10}	-9.8562×10^{-1}
y_{j4}	-7.1052×10^{-1}	-6.559×10^{-1}	4.148×10^{-3}	9.965×10^{-4}
y_{j1}^A	-2.9289	3.1738	$-7.8969 \times 10^{+3}$	8.663×10^{-1}
y_{j2}^A	-6.7467×10^{-3}	3.4307×10^{-6}	-9.953×10^{-1}	3.168×10^{-10}
y_{j3}^A	-5.036×10^{-1}	5.442×10^{-1}	-2.134×10^{-4}	-8.6605×10^{-1}
y_{j4}^A	-1.4014	7.102×10^{-4}	9.8568×10^{-1}	-4.6887×10^{-7}

Steady State 3

	1	2	3	4
y_{j1}	7.779×10^{-5}	1.978×10^{-2}	2.925×10^{-8}	1.597×10^{-1}
y_{j2}	-7.037×10^{-1}	-7.202×10^{-1}	-9.999×10^{-1}	-2.646×10^{-2}
y_{j3}	7.803×10^{-5}	1.907×10^{-2}	4.612×10^{-9}	-9.669×10^{-1}
y_{j4}	-7.105×10^{-1}	-6.933×10^{-1}	4.813×10^{-3}	1.973×10^{-1}
y_{j1}^A	$-4.248 \times 10^{+1}$	$4.377 \times 10^{+1}$	-1.652	8.598×10^{-1}
y_{j2}^A	-6.769×10^{-3}	2.807×10^{-5}	-9.953×10^{-1}	2.509×10^{-9}
y_{j3}^A	-7.305	7.232	-4.466×10^{-2}	-8.922×10^{-1}
y_{j4}^A	-1.406	5.559×10^{-3}	9.855×10^{-1}	-3.849×10^{-6}

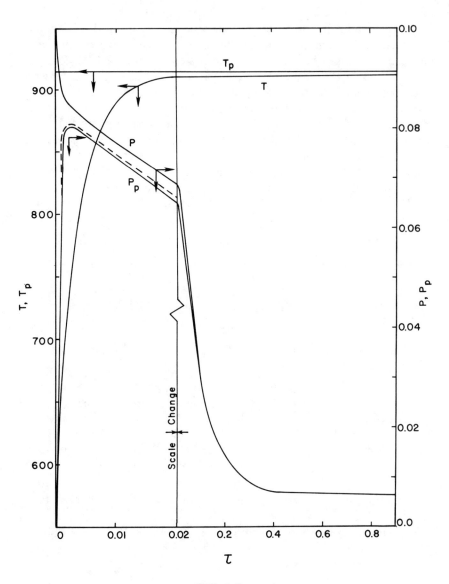

FIG. 9.1.
Approach to the high-temperature steady state. The dashed line is the reported numerical solution for P_p by Luss and Amundson. Note the change in scale on the time axis. (Fisher and Denn, 1978, copyright American Institute of Chemical Engineers, reproduced with permission.)

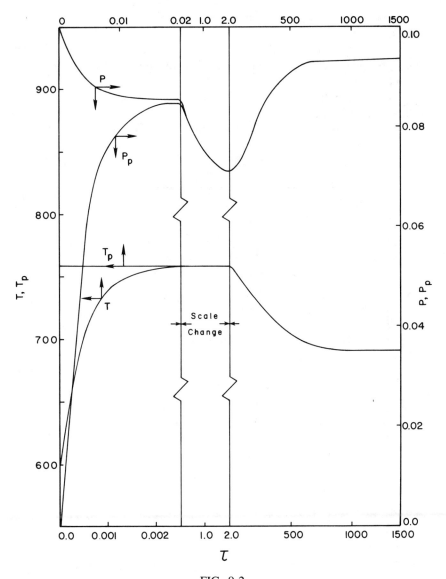

FIG. 9.2.
Approach to the low-temperature steady state. Note the two changes in scale on the time axis and the different initial time scales for temperatures and pressures. (Fisher and Denn, 1978, copyright American Institute of Chemical Engineers, reproduced with permission.)

in the figure. Note that the multiple time scales in the process are retained in the analytical solution.

Figure 9.2 shows results for a case in which the particles are initially 0.1 unit below the second (unstable) steady-state temperature. This case corresponds to Luss and Amundson's Fig. 4, and the analytical solution differs only slightly from their numerical calculation. The only deviation observable on this scale is in P_p; this is shown in Fig. 9.3, together with a 1974 numerical solution by Aiken and Lapidus using the highly accurate Gear method

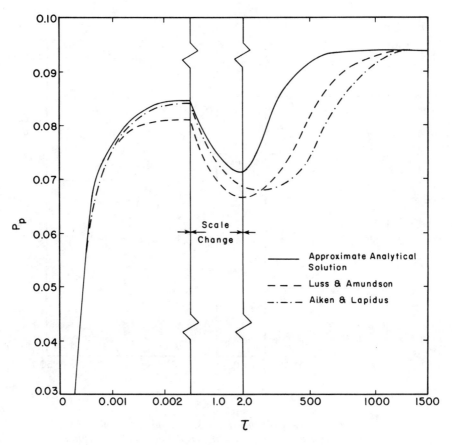

FIG. 9.3.
Particle temperature for conditions in Fig. 9.2; comparison of analytical solution to numerical results of Luss and Amundson, and Aiken and Lapidus. (Fisher and Denn, 1978, copyright American Institute of Chemical Engineers, reproduced with permission.)

for stiff systems of differential equations. The transient response in this example is quite slow, and we note that here, too, the multiple time scales have been properly accounted for by the approximate analytical solution.

The linearization limits computed from Eq. (9.53) give essentially the same results for gas and particle properties. About steady state one, we have $\lambda_1/I_{111} = -119.8$ and $|x_1(0)|, |x_3(0)| \ll 0.016, |x_2(0)|, |x_4(0)| \ll 85$. About steady state three, $\lambda_1/I_{111} = 14.3$, and the linearization limits are $|x_1(0)|, |x_3(0)| \ll 0.0011, |x_2(0)|, |x_4(0)| \ll 10$. The calculations in Figs. 9.1 and 9.2 are both initially within a region where nonlinear terms are important.

The stability estimate about the low-temperature steady state from Eq. (9.54) is

$$|2.09\xi_1 + 0.0048\xi_2 + 0.36\xi_3 + \xi_4| \leq 85.46 \tag{9.57}$$

Perturbations in the gas temperature (ξ_2) are unimportant relative to the solid temperature (ξ_4) because of the rapid equilibration, and pressure perturbations are also generally unimportant because of the relative magnitudes of temperatures and pressures. This stability estimate is consistent with the behavior of the full nonlinear system, in which the upper or lower steady state is approached depending on whether the solid temperature is above or below that of the unstable middle steady state. The estimate of the allowed perturbation is a bit high, however, since the middle steady state corresponds to a value of 68.82 for the linear combination in Eq. (9.57).

The stability estimate about the high temperature steady state is

$$|30.21\xi_1 + 0.0048\xi_2 + 5.20\xi_3 + \xi_4| \leq 10.18 \tag{9.58}$$

The dominant role of the particle temperature in determining stability is again made clear, though the estimate is a very conservative one. The intermediate steady state corresponds to a value of -154.53 for the linear combination in Eq. (9.58).

9.9 CONCLUDING REMARKS

Modal analysis can be a powerful tool, and it appears that it has been underutilized in the analysis of process models. The concept of decomposing a system into its independent dynamical modes is commonplace in structural analysis and control theory, but seems not to have established a central place in the mathematical framework of process engineering. This is a pity, for while the modal (or spectral) decomposition adds no information, it often provides insight leading to substantial simplification.

BIBLIOGRAPHICAL NOTES

Discussions of the basic elements of modal analysis can be found in

M. M. Denn, *Stability of Reaction and Transport Processes* (Englewood Cliffs, NJ: Prentice-Hall, Inc., 1975).

J. C. Friedly, *Dynamic Behavior of Processes* (England Cliffs, NJ: Prentice-Hall, Inc., 1972).

L. A. Gould, *Chemical Process Control: Theory and Applications* (Reading, MA: Addison-Wesley Publishing Co., 1969).

For a more recent discussion, focusing on the propagation of dynamical information through the modes and the implications with regard to model reduction, see

D. Bonvin and D. A. Mellichamp, *Int. J. Control* **35** (1982) 807, 829.

Section 9.7 is based on

E. H. Chimowitz, B. Greenbaum, D. R. Lewin, and M. M. Denn, "Control Structure for a Fluid Catalytic Cracker," unpublished paper presented at the AIChE Annual Meeting, Chicago, 1980.

References to the earlier work on the FCC are cited in Chapter 5.
Section 9.8 follows

R. J. Fisher and M. M. Denn, *AIChE J.* **24** (1978) 519.

The fluidized bed model is from

D. Luss and N. R. Amundson, *AIChE J.* **14** (1968) 211,

and the calculations by Aiken and Lapidus appear in

R. C. Aiken and L. Lapidus, *AIChE J.* **20** (1974) 368.

The basic development is in

W. Eckhaus, *Studies in Non-Linear Stability Theory* (New York, NY: Springer-Verlag, 1965).

10

DISCRETIZATION

10.1 INTRODUCTION

Models of physical phenomena are usually expressed in terms of ordinary and partial differential, integral, or integrodifferential equations. The *implementation* of a model in the form of a simulation usually requires a discrete approximation to the model equations.* The discretization often employs no more than standard methods for the numerical solution of algebraic or ordinary differential equations, usually through the use of well-documented library programs, and we presume that these are familiar to readers of this text. Some of the commonly used techniques, however, notably *methods of weighted residuals* and their offspring, *finite-element methods,* are grounded in analytical methods that are beyond the usual experience of traditionally educated engineers and scientists. We shall sketch out some of the basic concepts in this chapter, but only to the extent necessary to develop an approach to a class of problems; details that can lead to workable (and correct) computer codes must be sought in the appropriate numerical analysis literature.

10.2 METHODS OF WEIGHTED RESIDUALS

All methods of weighted residuals employ a common approach to the approximate solution of differential equations. It is convenient to start with a

---

*There is an unfortunate tendency to refer to *computer models* and *computer modeling*. The numerical solution is but one possible realization of a model, and the predictions of a simulation might fail to be those of the model to the extent that errors are introduced through the approximations inherent in numerical solutions. The model is the embodiment of the physics, not of the numerics.

linear example. Let $\mathscr{L}$ be a linear second-order differential operator (for example, $\mathscr{L}$ might be $d^2/dx^2 + x^3 d/dx + \tanh x$) with associated boundary conditions on the interval $0 < x < 1$:

$$\mathscr{L}[u] - f(x) = 0, \quad 0 < x < 1 \tag{10.1}$$

$$u(0) = u_0, \quad u(1) = u_1 \tag{10.2}$$

(Clearly $\mathscr{L}$ could be of higher order, and more general linear boundary conditions could also be used.) We then seek an approximate solution to Eq. (10.1) in the form

$$u(x) = \psi_0(x) + \sum_{n=1}^{N} C_n \psi_n(x) \tag{10.3}$$

$$\psi_0(0) = u_0, \quad \psi_0(1) = u_1 \tag{10.4a}$$

$$\psi_n(0) = \psi_n(1) = 0, \quad n = 1, 2, \ldots, N \tag{10.4b}$$

Substitution of Eq. (10.3) into Eq. (10.1) will not result in zero, because the solution is only approximate; we therefore have a position-dependent *residual*, $\mathscr{R}$, defined

$$\mathscr{R}(x) = \mathscr{L}\left[\psi_0 + \sum_{n=1}^{N} C_n \psi_n\right] - f(x)$$

$$= \mathscr{L}[\psi_0] + \sum_{n=1}^{N} C_n \mathscr{L}[\psi_n] - f(x) \neq 0 \tag{10.5}$$

The second equality follows from the linearity of the operator $\mathscr{L}$.

We now seek to solve Eq. (10.1) *on the average* by selection of the values of the coefficients $\{C_n\}$; we choose N weighting functions $\{w_n(x)\}$ and set N weighted averages of the residual to zero:

$$\int_0^1 w_n(x) \mathscr{R}(x)\, dx = \int_0^1 w_n(x) \mathscr{L}[\psi_0(x)]\, dx + \sum_{n=1}^{N} C_n \int_0^1 w_n(x) \mathscr{L}[\psi_n(x)]\, dx$$

$$- \int_0^1 w_n(x) f(x)\, dx = 0, \quad n = 1, 2, \ldots N \tag{10.6}$$

Equation (10.6) gives N linear equations for the N unknowns, $\{C_n\}$, and hence the solution.

The success of the method obviously depends on the judicious choice

10.2 METHODS OF WEIGHTED RESIDUALS

both of the approximating functions, $\{\psi_n(x)\}$, and the weighting functions, $\{w_n(x)\}$. The former must clearly be linearly independent and should have a shape appropriate to the anticipated solution; orthogonal functions are often used. The selection of the weighting functions is less clear, and different choices lead to the several popular methods of weighted residuals.

In the method of *collocation*, the weighting functions are chosen equal to Dirac delta functions:

$$w_n(x) = \delta(x - x_n), \quad n = 1, 2, \ldots, N \tag{10.7}$$

This is equivalent to setting the residual to zero at N finite points. *Orthogonal collocation* is usually employed, where the collocation points $\{x_n\}$ are the zeros of an orthogonal polynomial. The basic concept is clear, though actual implementation requires consideration of the proper choice of orthogonal polynomials. The calculations in Section 7.4.5 were carried out using orthogonal collocation after first dividing the radius into two segments, with matching boundary conditions at the intersection of the two segments; Jacobi polynomials were used on the inner element and shifted Legendre polynomials on the outer, with the choice dictated by the boundary conditions.

The *Galerkin* method is perhaps the most widely used, because it forms the basis of most finite-element techniques. In this method the weighting functions and the trial functions are the same:

$$w_n(x) = \psi_n(x), \quad n = 1, 2, \ldots, N \tag{10.8}$$

This choice is not at all transparent, and is the focus of much of this chapter.

The differential equation need not be linear. In that case the coefficients $\{C_n\}$ must be found by solution of N nonlinear algebraic equations, usually using Newton's method. As an illustration, consider the equation

$$\mathcal{L}[u] + g(u) - f(x) = 0, \quad 0 < x < 1 \tag{10.9}$$

with boundary conditions like those in (10.2), but as many as are appropriate to the order of the operator; $g(u)$ is a differentiable nonlinear function, while $\mathcal{L}$ remains a linear operator. Let $\{\bar{C}_n\}$ denote an estimate of the coefficients in an approximate solution, and define

$$\bar{u}(x) = \psi_0(x) + \sum_{n=1}^{N} \bar{C}_n \psi_n(x) \tag{10.10}$$

The next estimate,

$$u(x) = \bar{u}(x) + \delta u(x) \tag{10.11}$$

is assumed to satisfy Eq. (10.9). We therefore have

$$\mathcal{L}[\bar{u} + \delta u] + g(\bar{u} + \delta u) - f(x) = 0 \tag{10.12}$$

Using the linearity of the differential operator, and expanding $g(u)$ in a Taylor series about $\bar{u}$ and retaining only the first-order term, we then obtain

$$\mathcal{L}[\delta u] + g'(\bar{u})\delta u + \{\mathcal{L}[\bar{u}] + g(\bar{u}) - f(x)\} = 0 \tag{10.13}$$

$g'(\bar{u})$ is dg/du evaluated at $u = \bar{u}$. We can now solve this *linear* equation by any method; we choose the method of weighted residuals, thus obtaining corrections $\{\delta C_n\}$ to the coefficients $\{\bar{C}_n\}$:

$$\delta u(x) = \sum_{n=1}^{N} \delta C_n \psi_n(x) \tag{10.14}$$

The usual procedure then leads to the following set of linear equations for the corrections $\{\delta C_n\}$:

$$\sum_{n=1}^{N} A_{mn} \delta C_n = \int_0^1 w_m(x)\{\mathcal{L}[\bar{u}] + g(\bar{u}) - f(x)\}dx \tag{10.15a}$$

$$A_{mn} = \int_0^1 w_m(x)\{\mathcal{L}[\psi_n] + g'(\bar{u})\psi_n\}dx \tag{10.15b}$$

Note that the right-hand side of Eq. (10.15a) equals zero when $\bar{u}(x)$ is such that all weighted residuals of Eq. (10.9) equal zero. (One arrives at the same final iterative equation by linearizing after setting the weighted residuals of the nonlinear equation to zero, and that is the more common approach. We have chosen this sequence in order to emphasize that the iterative solution of the nonlinear differential equation through successive solution of a set of linear equations—in this case Eq. (10.13)—is a general procedure that does not depend on the use of a particular scheme like weighted residuals.)

While we have presented the methods of weighted residuals here for equations with a single independent variable, the method generalizes in two ways. Suppose we have two independent variables, x and y; we can then seek approximate solutions in the form

$$u(x,y) = \Psi_0(x,y) + \sum_{n,m=1}^{N} C_{nm}\psi_n(x)\phi_m(y) \tag{10.16}$$

where $\Psi_0(x,y)$ satisfies all nonhomogeneous boundary conditions, and $\{\psi_n\}$,

$\{\phi_m\}$ are specified functions with homogeneous boundary conditions. This is the approach used in finite-element methods for partial differential equations. Alternatively, we might seek solutions in the form

$$u(x,y) = \Psi_0(x,y) + \sum_{n=1}^{N} \phi_n(y)\psi_n(x) \qquad (10.17)$$

where the $\{\phi_n(y)\}$ are functions to be determined. Weighting is carried out only over x, and the result is a set of ordinary differential equations for the $\{\phi_n(y)\}$. This is the approach used in the coal gasifier calculations in Section 7.4.5.

10.3 THE ADJOINT OPERATOR

We employed the concept of the adjoint in Section 9.3 for a linear algebraic operator. We need to introduce the adjoint operator here in order to set the stage for the development of the Galerkin method. Consider the linear operator $\mathcal{L}[u(x)]$, defined over $0 < x < 1$, with linear homogeneous boundary conditions on u at $x = 0, 1$. We now define the adjoint operator $\mathcal{L}^A$ and the adjoint function $u^A(x)$ such that

$$\int_0^1 u^A(x)\mathcal{L}[u(x)]dx = \int_0^1 u(x)\mathcal{L}^A[u^A(x)]dx \qquad (10.18)$$

Equation (10.18) defines the linear boundary conditions on $u^A(x)$.

The adjoint operator has some useful properties, but we shall establish only one: *The eigenvalues of $\mathcal{L}$ and $\mathcal{L}^A$ are equal, and the eigenfunctions are orthorgonal*. Consider the two eigenvalue problems

$$\mathcal{L}[u(x)] = \lambda u(x) \qquad (10.19a)$$
$$\mathcal{L}^A[u^A(x)] = \lambda^A u^A(x) \qquad (10.19b)$$

with the associated linear homogeneous boundary conditions. We now multiply Eq. (10.19a) by $u^A(x)$, Eq. (10.19b) by $u(x)$, integrate from zero to unity, and subtract; from Eq. (10.18) we have

$$(\lambda - \lambda^A)\int_0^1 u(x)u^A(x)dx = 0 \qquad (10.20)$$

from which it follows that $\lambda = \lambda^A$. It further follows that eigenfunctions corresponding to different eigenvalues are orthogonal; i.e., let $u_n(x)$, $u_n^A(x)$ be the eigenfunctions corresponding to λ_n; then

$$\int_0^1 u_n(x) u_m^A(x) dx = 0, \quad n \neq m \tag{10.21}$$

These are equivalent to conditions used for eigenvectors in the modal expansions in Section 9.3.

A *self-adjoint operator* is equal to its own adjoint; i.e.,

$$\int_0^1 v(x) \mathcal{L}[u(x)] dx = \int_0^1 u(x) \mathcal{L}[v(x)] dx \tag{10.22}$$

for any differentiable $u(x)$, $v(x)$ satisfying the boundary conditions for $\mathcal{L}[u]$. It is straightforward to establish that the eigenvalues of a self-adjoint operator are real, as follows:

Let $\lambda = \lambda_R + i\lambda_I$, $u(x) = u_R + iu_I$, where i is the imaginary operator ($i^2 = -1$). Then

$$\mathcal{L}[u_R + iu_I] = \mathcal{L}[u_R] + i\mathcal{L}[u_I] = (\lambda_R + i\lambda_I)(u_R + iu_I)$$
$$= \lambda_R u_R - \lambda_I u_I + i(\lambda_R u_I + \lambda_I u_R) \tag{10.23}$$

Equating real and imaginary parts of this complex equation, we have

$$\mathcal{L}[u_R] = \lambda_R u_R - \lambda_I u_I \tag{10.24a}$$

$$\mathcal{L}[u_I] = \lambda_R u_I + \lambda_I u_R \tag{10.24b}$$

We now multiply the first of this pair by u_I, the second by u_R, integrate from zero to unity, and subtract; since $\int u_R \mathcal{L}[u_I] dx = \int u_I \mathcal{L}[u_R] dx$, we obtain

$$\lambda_I \int_0^1 (u_R^2 + u_I^2) dx = 0 \tag{10.25}$$

from which it follows that $\lambda_I = 0$ and λ is real. Thus, the eigenvalues of a self-adjoint operator can be ordered, $\lambda_1 > \lambda_2 > \lambda_3 > \cdots$. Since $u_n^A(x) = u_n(x)$ it follows directly from Eq. (10.21) that the eigenfunctions of a self-adjoint operator are orthogonal.

The differential operators that commonly arise in problems in mathematical physics are often self-adjoint, and these properties are important with regard to ease of solution. The convergence of certain approximation schemes is monotonic when the operator is self-adjoint, making it possible

10.4 CALCULUS OF VARIATIONS

to establish bounds on the approximation. It can be shown that all linear second-order ordinary differential operators with homogeneous boundary conditions can be transformed to the self-adjoint operator

$$\mathcal{L}[u] = \frac{d}{dx} p(x) \frac{du}{dx} + q(x)u \qquad (10.26a)$$

$$a_0 u(0) + b_0 u'(0) = 0, \qquad a_1 u(1) + b_1 u'(1) = 0 \qquad (10.26b)$$

Such a general result is not available for higher-order systems.

10.4 CALCULUS OF VARIATIONS

The calculus of variations is a powerful tool in many areas of analysis. We need one specific result, but we shall first introduce the subject with the most basic problem. Consider the integral

$$\mathcal{E}[u] = \int_a^b \mathcal{F}(u, u', x) dx \qquad (10.27)$$

where $u(x)$ is a continuously differentiable function satisfying certain boundary conditions at $x = a$ and b; $u' = du/dx$; and $\mathcal{F}$ is continuously differentiable with respect to the first two arguments. For each choice of function $u(x)$ satisfying the prescribed boundary conditions the *functional* $\mathcal{E}[u]$ takes on a numerical value; we seek the function $u_m(x)$ that makes $\mathcal{E}$ a minimum.

There are several equivalent approaches. Let ϵ be an arbitrary but small real number, and let $\eta(x)$ be an arbitrary continuously differentiable function that satisfies homogeneous boundary conditions at $x = a$ and b. We choose

$$u(x) = u_m(x) + \epsilon \eta(x) \qquad (10.28)$$

(The particular homogeneous boundary conditions on η are dictated by the fact that u and u_m must both satisfy the specified conditions on u at $x = a, b$.) For given u_m and η, $\mathcal{E}$ is an ordinary function of ϵ:

$$\mathcal{E}(\epsilon) = \int_a^b \mathcal{F}(u_m(x) + \epsilon \eta(x), u_m'(x) + \epsilon \eta'(x), x) dx \qquad (10.29)$$

Since $u_m(x)$ is the function that causes $\mathcal{L}$ to take on its minimum value, the minimum with respect to ϵ must occur at $\epsilon = 0$; thus, $d\mathcal{E}/d\epsilon$ must vanish at $\epsilon = 0$:

$$\left.\frac{d\mathcal{E}}{d\epsilon}\right|_{\epsilon=0} = \int_a^b \left\{ \left.\frac{\partial \mathcal{F}}{\partial u}\right|_{u=u_m} \eta + \left.\frac{\partial \mathcal{F}}{\partial u'}\right|_{u=u_m} \eta' \right\} dx = 0 \qquad (10.30)$$

Henceforth we will delete the notation "$u = u_m$" with the understanding that $\mathcal{F}$ and its derivative are to be evaluated at the minimizing function.

We now integrate the second term by parts to obtain

$$\int_a^b \left\{ \frac{\partial \mathcal{F}}{\partial u} - \frac{d}{dx}\frac{\partial \mathcal{F}}{\partial u'} \right\} \eta(x) dx + \frac{\partial \mathcal{F}(b)}{\partial u'} \eta(b) - \frac{\partial \mathcal{F}(a)}{\partial u'} \eta(a) = 0 \qquad (10.31)$$

The easiest case to deal with is one in which $u(a)$ and $u(b)$ are specified; then $\eta(a) = \eta(b) = 0$ and the final two terms in Eq. (10.31) equal zero. Since $\eta(x)$ is arbitrary in the interval (a,b) we may choose any continuously differentiable function, and the selection that facilitates reaching the solution we seek is

$$\eta(x) = w(x) \left\{ \frac{\partial \mathcal{F}}{\partial u} - \frac{d}{dx}\frac{\partial \mathcal{F}}{\partial u'} \right\}, \quad w(x) > 0 \text{ in } a < x < b \qquad (10.32)$$

Equation (10.31) then becomes

$$\int_a^b w(x) \left\{ \frac{\partial \mathcal{F}}{\partial u} - \frac{d}{dx}\frac{\partial \mathcal{F}}{\partial u'} \right\}^2 dx = 0 \qquad (10.33)$$

The integral of a nonnegative function over a positive interval can equal zero only if the integral is zero, so we obtain the *Euler equation:*

$$\frac{\partial \mathcal{F}}{\partial u} - \frac{d}{dx}\frac{\partial \mathcal{F}}{\partial u'} = 0 \qquad (10.34)$$

Equation (10.34) is a second-order, ordinary differential equation that must be satisfied by the minimizing function $u_m(x)$ in the interval $a < x < b$. The fact that it is second order is best illustrated by making further differentiability assumptions on $\mathcal{F}$ and using the chain rule to obtain

$$\frac{\partial \mathcal{F}}{\partial u} - \frac{\partial^2 \mathcal{F}}{\partial x \partial u'} - \frac{\partial^2 \mathcal{F}}{\partial u \partial u'} u' - \frac{\partial^2 \mathcal{F}}{(\partial u')^2} u'' = 0 \qquad (10.35)$$

We now return to Eq. (10.31). It should be clear that the logic leading

10.4 CALCULUS OF VARIATIONS

to the Euler equation is in no way dependent on the vanishing of $\eta(x)$ at the endpoints, since the integral must still equal zero for arbitrary η. If $u(a)$ is not specified, then $\eta(a)$ is arbitrary; in that case we can choose $\eta(a)$ proportional to $\partial \mathcal{F}(a)/\partial u'$, leaving us with a square that is equal to zero, and similarly at $x = b$. Carrying this logic through, we obtain the following boundary conditions for Eq. (10.34):

$$\text{either} \quad u \text{ is specified at } x = a, b \quad (10.36a)$$

$$\text{or} \quad \frac{\partial \mathcal{F}}{\partial u'} = 0 \quad (10.36b)$$

The condition (10.36b) is known in the calculus of variations and numerical analysis literature as a *natural boundary condition,* and it plays an important role in finite element analysis.

We can now pose the question: *Under what conditions is a differential equation the Euler equation for a problem in the calculus of variations?* There is no general solution to this *inverse problem,* but we will show that the answer is always affirmative if the differential operator is self-adjoint. Since every linear second-order ordinary differential equation can be transformed to a self-adjoint form, this then means that all linear second-order ordinary differential equations are Euler equations for some problem in the calculus of variations; it can be shown that this statement is not true for all third-order equations. We should also note that, while we formulated the original problem as one of minimizing an integral, we have in fact only used the vanishing of the first derivative and are therefore really only talking about *stationary* values (maximum, minimum, or saddle); further results are required to establish the conditions for a minimum.

Consider now the problem of finding the stationary conditions for

$$\mathcal{E}[u] = \int_0^1 \{\tfrac{1}{2} u \mathcal{L}[u] + G(u) - f(x)u\} dx \quad (10.37)$$

where $\mathcal{L}[u]$ is a self-adjoint operator with associated linear boundary conditions. We again set $u(x) = u_m(x) + \epsilon \eta(x)$, differentiate with respect to ϵ, and set ϵ to zero to obtain

$$\int_0^1 \{\tfrac{1}{2} \eta \mathcal{L}[u_m] + \tfrac{1}{2} u_m \mathcal{L}[\eta] + g(u_m)\eta - f(x)\eta\} dx = 0 \quad (10.38)$$

where $g(u) = dG/du$. Because the operator $\mathcal{L}[\ ]$ is self-adjoint, we have

$\int u_m \mathcal{L}[\eta]dx = \int \eta \mathcal{L}[u_m]dx$, and we may write Eq. (10.37) as

$$\int_a^b \{\mathcal{L}[u_m] + g(u_m) - f(x)\}\eta(x)dx = 0 \qquad (10.39)$$

Because of the arbitrariness of $\eta(x)$ we are then led to the vanishing of the term in braces, which establishes Eq. (10.9) as the Euler equation for the problem Eq. (10.37).

10.5 RITZ-GALERKIN METHOD

Problems in the calculus of variations are solved traditionally by either of two general approaches: *indirect methods,* in which the equations establishing necessary conditions for an extremum (the Euler equation, or appropriate generalization) are solved; and *direct methods,* in which the minimization problem is attacked directly without the intermediate step of obtaining necessary conditions. The latter can provide an efficient means of solving a differential equation through a direct solution of the equivalent variational problem.

The Ritz method is an approximate procedure for obtaining direct solutions to problems in the calculus of variations by reduction to a problem in multivariable calculus. The minimizing function is approximated by a functional form having some undetermined coefficients, and the integral is evaluated. The resulting function is then minimized with respect to the coefficients. An interesting result is obtained for variational problems in which the Euler equation is self-adjoint, as follows:

Consider the problem defined by Eq. (10.37). We will suppose that we seek an approximate solution in the form of Eq. (10.4). We can then write

$$\mathcal{E} = \int_0^1 \left\{ \frac{1}{2}\left(\psi_0 + \sum_{n=1}^N C_n\psi_n\right)\left(\mathcal{L}[\psi_0] + \sum_{m=1}^N C_m\mathcal{L}[\psi_m]\right) \right. \\ \left. + G\left(\psi_0 + \sum_{n=1}^N C_n\psi_n\right) - f(x)\left(\psi_0 + \sum_{n=1}^N C_n\psi_n\right)\right\}dx \qquad (10.40)$$

In the usual application of the Ritz method we would now carry out the indicated integrations. Instead, we will immediately seek the stationary points of $\mathcal{E}$ with respect to the $\{C_n\}$ by setting derivatives to zero:

$$\frac{\partial \mathscr{E}}{\partial C_k} = \int_0^1 \left\{ \frac{1}{2} \psi_k \left(\mathscr{L}[\psi_0] + \sum_{n=1}^{N} C_n \mathscr{L}[\psi_n] \right) \right.$$

$$+ \frac{1}{2} \left(\psi_0 + \sum_{m=1}^{N} C_m \psi_m \right) \mathscr{L}[\psi_k]$$

$$\left. + g\left(\psi_0 + \sum_{n=1}^{N} C_n \psi_n \right) \psi_k - f(x)\psi_k \right\} = 0,$$

$$k = 1, 2, \ldots, N \qquad (10.41)$$

Recall that $g = dG/du$. Because the operator $\mathscr{L}$ is self-adjoint, we replace $\int \psi_m \mathscr{L}[\psi_k]dx$ by $\int \psi_k \mathscr{L}[\psi_m]dx$, $m = 0, 1, 2, \ldots, N$, in which case the first two terms in the integral become identical. We therefore have

$$\int_0^1 \{\mathscr{L}[\bar{u}] + g(\bar{u}) - f(x)\} \psi_k(x) dx = 0, \quad k = 1, 2, \ldots, N \qquad (10.42)$$

where $\bar{u}$ denotes the approximate solution. The term in braces is simply the residual of Eq. (10.9) with the approximation (10.3), so we have recovered the Galerkin method of weighted residuals, for which the weighting functions are identical to the approximating functions.

Equation (10.42) provides the theoretical foundation for the Galerkin method when it is applied to self-adjoint systems; while it is a nonobvious procedure for obtaining approximate solutions to differential equations, it is identical to what is perhaps the most obvious and intuitively appealing method of obtaining a direct solution to the equivalent variational problem. Quite strong results regarding size convergence (number of terms in the approximation), monotonicity, and quality of approximation are available because of the self-adjoint nature of the system.

Many differential equations encountered in practice are not self-adjoint, unfortunately. The Galerkin method of weighted residuals may still be used, but for these systems the choice of weighting functions lacks the fundamental basis that stems from the connection with the calculus of variations. It remains, nevertheless, one of the most frequently employed methods of discrete approximation.

10.6 FINITE-ELEMENT METHODS

In finite-element methods the spatial region over which the equations are to be solved is broken up into smaller elements, usually quadrilaterals or triangles; these elements need not be of uniform size, so complex shapes can be accommodated and the element size can be based on the expected rapidity of change of the computed functions. The solution is approximated *over each element* using a method of weighted residuals, usually the Galerkin method; because only a portion of the total solution is being approximated on each element, low-order polynomials—usually linear or quadratic—can be used with sufficient accuracy as approximating functions. The requirement that the approximate solution computed on adjacent elements be the same where the elements meet provides additional equations relating the coefficients of the approximation on adjacent elements, and these equations must be solved together with those resulting from setting the weighted residuals to zero.

This basic idea has been widely implemented, particularly in applications in mechanics, and a large number of quite general computer codes exist, many with automatic grid-generating capabilities. Large numbers of equations for the coefficients must be solved simultaneously; since the equations are usually nonlinear, an iterative scheme like that in Eq. (10.15) is generally required. The matrices that must be inverted will consist of dense blocks associated with the individual elements, with sparse coupling because of the continuity requirements on adjacent elements. The application of this method to the upper portion of a fiber spinline is described in Section 12.4. It is important to emphasize that the basic idea is quite straightforward, but implementation requires considerable care in order to develop a logical structure that can be coded for a particular set of equations and approximating functions in such a way that the number and location of elements can be changed without any program revision.

10.7 CONCLUDING REMARKS

Our general approach has been to assume a readership that is familiar with basic numerical tools for the solution of algebraic and differential equations. As a result, this chapter on discretization has evolved as a most specialized one, which perhaps bears directly on only a small fraction of the modeling problems that might be encountered. The topics that have been covered (in the most sketchy manner, it must be conceded) represent those that we have found to be missing from the common experience of engineers and scientists, though such *should* not be the case. They are particularly relevant to an understanding of the modern numerical methods that will inevitably enter into the application of most models.

BIBLIOGRAPHICAL NOTES

Two excellent introductions to methods of approximation, with many examples, are

> B. A. Finlayson, *The Method of Weighted Residuals and Variational Methods* (New York, NY: Academic Press, 1972).
>
> J. Villadsen and M. L. Michelsen, *Solution of Differential Equations by Polynomial Approximation* (Englewood Cliffs, NJ: Prentice-Hall, 1978).

The standard textbook on the calculus of variations is

> G. A. Bliss, *Lectures on the Calculus of Variations* (Chicago, IL: University of Chicago Press, 1946).

Our own personal favorites are

> N. I. Akhiezer, *The Calculus of Variations* (New York, NY: Blaisdell Publishing Co., 1962).
>
> L. A. Pars, *An Introduction to the Calculus of Variations* (London: Heinemann Educational Books, 1962).

The subject is covered in all its complexities, but in terms of a few specific problems, in a wonderfully readable monograph by Bliss:

> G. A. Bliss, *Calculus of Variations* (LaSalle, IL: Open Court Publishing Co., 1925).

For a very different point of view, with emphasis on nonclassical applications, see

> M. M. Denn, *Optimization by Variational Methods* (New York, NY: McGraw-Hill, 1969; reprinted with corrections and supplementary bibliography, New York, NY: Robert Krieger, 1978).

In addition to the examples in Finlayson and Villadsen and Michelsen, numerous applications of Galerkin's method to problems in stability are found in

> M. M. Denn, *Stability of Reaction and Transport Processes* (Englewood Cliffs, NJ: Prentice-Hall, 1975).

Finite element methods are discussed in many recent texts, such as

> E. B. Becker, G. F. Carey, and J. T. Oden, *Finite Elements: An Introduction* (Englewood Cliffs, NJ: Prentice-Hall, 1981).

M. F. Yeo and Y. K. Cheung, *A Practical Introduction to Finite Element Analysis* (London: Pitman, 1979).

O. C. Zienkiewicz, The Finite Element Method, 3rd edition (London: McGraw-Hill (U.K.), 1977).

The application to flows of polymeric liquids, which is touched upon in Chapter 12, introduces special problems; see

M. J. Crochet, A. R. Davies, and K. Walters, *Numerical Solution of Non-Newtonian Flow* (Amsterdam: Elsevier, 1984).

11

FROM DISCRETE TO CONTINUOUS

11.1 INTRODUCTION

Process models are usually ordinary or partial differential equations, or perhaps integrodifferential equations. The mathematical structure is inherently continuous in the independent variables. Numerical computation with such a model usually involves an explicit or implicit discretization, as in the use of numerical integration methods, weighted residual methods, etc.

Some problems are inherently discrete, either because of a spatial staging or a cascading that involves only integral values of a process variable. The former case is best illustrated by a separation process like distillation, in which there is flow between a large number of identical stages, with concentration changes occurring at the discrete spatial locations of the stages. The latter is illustrated by a polymerization reactor, in which the average number of repeat units in a molecule changes by integral values.

Discrete problems are usually studied (when an analytical treatment is possible) by means of the calculus of finite differences. There are times, however, when it is convenient to reverse the usual procedure and to develop a *continuous* approximation to the discrete model. This approach might be familiar from the introduction to quantum mechanics in a basic physics or physical chemistry course, where the discrete energy levels of the hydrogen atom are approximated as a continuous function. The application to distillation that we shall illustrate here was first utilized in the U.S. Manhattan

Project of World War II in the analysis of the gaseous diffusion process for the separation of uranium isotopes.

11.2   BINARY DISTILLATION

Distillation is a process in which a volatile component is separated from a mixture by a succession of vaporization and condensation steps. We restrict ourselves here to binary (two-component) mixtures. Let* x be the mole fraction of the more volatile species in a liquid mixture and y the mole fraction in the vapor. At a fixed temperature and pressure the system will equilibrate, with a unique relationship between the composition of the phases:

$$y = \phi(x) \tag{11.1}$$

This is the important constitutive equation for the process. (Temperature appears implicitly in this constitutive equation at a given system pressure, since y is uniquely fixed for each temperature and pressure.) The equilibrium function for the ethanol-water system (which is highly nonideal) is shown in Fig. 11.1. If the vapor is separated from the liquid and condensed, the condensate will be richer in the volatile component than the original mixture as long as $y > x$.

A tray distillation column is shown schematically in Fig. 11.2. This column is a means of continuously equilibrating and separating mixtures in order to effect a separation. It is an energy-intensive process. Heat added to the bottom of the column forms a vapor that rises. Vapor leaving the top is cooled to form a liquid that flows downward. There is a liquid holdup on each tray. Vapor rising from the tray below enters the tray (usually through a series of perforations) and mixes with the liquid. The liquid flows over a weir to the tray below and is replenished from the tray above. There is energy transfer between the liquid and vapor such that some vapor is condensed and some liquid is vaporized. The vapor leaving the tray is assumed to have been in contact with the liquid for a sufficiently long time to reach equilibrium (or, more commonly, to reach a certain fraction of the equilibrium composition based on the tray *efficiency*.)

Let H_n denote the total moles of liquid on the nth tray, and let L_n and V_n denote the respective molar flow rates of liquid and vapor leaving

---

*It is traditional in the mass transfer and separation processes literature to use x and y to denote mole fractions, and we follow that practice to facilitate comparison with that specialized literature. x and y are thus *dependent* variables, and not independent coordinate positions.

11.2 BINARY DISTILLATION

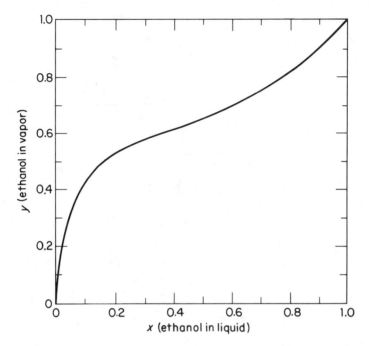

FIG. 11.1
Vapor-liquid equilibrium function for the ethanol-water system.

the nth tray. If we assume that there is no holdup in the vapor space, so that what leaves one tray is identical to what enters the tray above or below at the same instant, then the overall mass balance on a control volume encompassing the liquid and vapor above the nth tray and below tray $(n + 1)$ is

$$\frac{dH_n}{dt} = V_{n+1} + L_{n+1} - V_n - L_n \qquad (11.2)$$

Similarly, taking the liquid on the nth tray to be well mixed, the mass balance on the volatile component on the nth tray is

$$\frac{dx_n H_n}{dt} = V_{n-1}y_{n-1} + L_{n+1}x_{x+1} - V_n y_n - L_n x_n \qquad (11.3)$$

The energy equation should also be written for the control volume, but a simplification that is a useful first approximation obviates the need. Pure substances having similar boiling points tend to have similar enthalpy

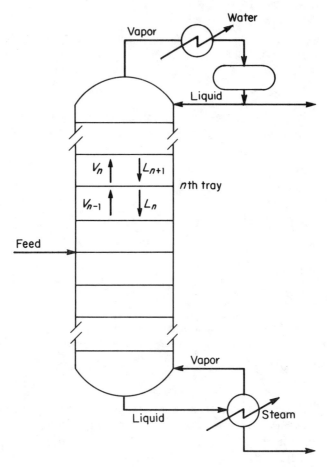

FIG. 11.2
Schematic of a tray distillation column.

changes on vaporization. (The molar enthalpies of vaporization of ethanol and water, for example, are 9.4 and 9.5 Kcal, respectively.) Thus, the enthalpy change resulting from condensation or vaporization of one mole of the mixture is essentially the same, regardless of composition. If the process is also assumed to be adiabatic, then all of the enthalpy change resulting from condensation goes into vaporization of an equal number of moles, and the energy balance leads directly to the approximation of constant vapor molar flow rate:

$$V_n = V_{n-1} = V = \text{constant} \tag{11.4}$$

It is convenient for our purposes to assume further that the time dependence of tray hydraulics can be neglected, and that liquid holdup on each tray is constant in time. In that case, $dH_n/dt = 0$, and together with Eq. (11.4) we obtain constant liquid molar flow rate:

$$L_n = L_{n+1} = L = \text{constant} \tag{11.5}$$

It is important to keep in mind that these approximations need not be made, and are not made in most models of real distillation towers, but their use helps to bring out important qualitative features that might be masked by a more complete treatment.

We assume that the vapor and liquid at each stage are in equilibrium: $y_n = \phi(x_n)$, or that Eq. (11.1) already includes the efficiency of the tray. Equations (11.1) through (11.5) then combine to form the single set of difference-differential equations

$$\frac{H}{L}\frac{dx_n}{dt} = x_{n+1} - \left[\frac{V}{L}\phi(x_n) + x_n\right] + \frac{V}{L}\phi(x_{n-1}) \tag{11.6}$$

V and L will be different above and below the feed point, and different equations must apply to the boiling stage at the bottom and to the condensor at the top, but we need not worry about these (very important) details for our purposes here.

11.3 CONTINUOUS APPROXIMATION

It is suggestive to rewrite Eq. (11.6) in the following way by adding and subtracting terms:

$$\frac{H}{L}\frac{dx_n}{dt} = [x_{n+1} - 2x_n + x_{n-1}] - \frac{V}{L}[\phi(x_n) - \phi(x_{n-1})] + [x_n - x_{n-1}] \tag{11.7}$$

Each term in brackets has the appearance of the numerator of the finite-difference approximation to a derivative. We can formalize the process somewhat by defining a continuous variable v, which ranges from zero to N, where N is the total number of trays. Define

$$z = v/N \tag{11.8}$$

We can think of x as a continuous function of z, with a physical meaning only when v takes on integral values and z has the meaning of the fractional number of trays up from the bottom of the column.

We now write $x = x(z,t)$, recalling that $x_n(t) = x(n/N, t)$. If we think of z as being fixed at some particular, but arbitrary, value of n, then we can rewrite Eq. (11.7) as

$$\frac{H}{L}\frac{\partial x}{\partial t}\bigg)_z = \frac{1}{N^2}\left[\frac{x(z + 1/N, t) - 2x(z,t) + x(z - 1/N, t)}{(1/N)^2}\right] \\ - \frac{V}{NL}\left[\frac{\phi(x(z,t)) - \phi(x(z - 1/N, t))}{(1/N)}\right] \\ + \frac{1}{N}\left[\frac{x(z,t) - x(z - 1/N, t)}{(1/N)}\right] \quad (11.9)$$

If N is very large, then it is reasonable to think in terms of the limit as $1/N \to 0$ with regard to the difference quotients. The first bracketed term in this limit is $\partial^2 x/\partial z^2$, while the second and third are $\partial \phi/\partial z$ and $\partial x/\partial z$, respectively. We can then write Eq. (11.9) in the limit as

$$\frac{N^2 H}{L}\frac{\partial x}{\partial t} = \frac{\partial^2 x}{\partial z^2} + N\left[1 - \frac{V}{L}\frac{d\phi}{dx}\right]\frac{\partial x}{\partial z}, \quad 0 \le z \le 1 \quad (11.10)$$

where the solution has physical meaning at the discrete values $z = n/N$. As before, two spatial conditions must be specified for the problem to be well posed.

The continuous approximation reveals features of the model structure that are much less obvious in the discrete formulation. The dynamics of the tower are like those of a transient system in which there is both diffusion (the $\partial^2 x/\partial z^2$ term) and convective flow (the $\partial x/\partial z$ term). The coefficient $(V/L)\, d\phi/dx - 1$ plays the role (to within a constant) of a spatially dependent convective velocity.

It is a characteristic of difficult separations [where $\phi(x)$ is close to x] that $(V/L)\, d\phi/dx$ is close to unity. If we take the value to be identically one, then Eq. (11.10) reduces to the diffusion equation. We see, therefore, that disturbances "diffuse" through the column. We know from the discussion on scaling in Chapter 6 that the characteristic response time will be of the order of $N^2 H/L$; when an exact solution is obtained for the appropriate boundary conditions, the time constant is in fact $N^2 H/6L$. The response time is therefore proportional to the square of the number of trays, and a very sluggish response is to be expected (and is indeed observed) in large columns.

11.4 BATCH POLYMERIZATION

There is no difficulty in principle in obtaining solutions for the more general case including the convection term. The equation is linear when ϕ is a linear function of x, which is often a good approximation. Linearized dynamics for small disturbances about the steady state will be described by an equation with the structure of Eq. (11.10), with position-dependent coefficients that depend on the steady-state composition profile. Pigford has given solutions to these equations, using the Rayleigh-Ritz method, and we shall not be concerned with the details. The "velocity" term $(V/L)\,d\phi/dx - 1$ induces a "convective flow" toward the end of the column from the interior, but reverses sign before the end because of the nonlinearity of Eq. (11.1) and causes a "flow" *into* the column. Thus, transient disturbances are "trapped" and cannot flow out of the column, but can only escape by the much slower diffusive mechanism.

The approach of approximating the discrete process by a continuous one undoubtedly has potential application in process control beyond the ease of identifying the dominant eigenvalues and modes. One of the basic questions in distillation column control is where to make measurements within the column to ensure maximum sensitivity. The continuous representation allows the use of approximation methods like orthogonal collocation or finite elements that exploit the known properties of the partial differential equation, without regard to the underlying stage-wise structure. The collocation or nodal points can therefore be chosen first, and their locations will then dictate the placement of sensors. (This approach does not require the continuous approximation, but it is greatly facilitated by the extensive experience on solution methods for diffusion-convection systems.)

11.4 BATCH POLYMERIZATION

Polymers are macromolecules made up of a large number of monomer repeat units. Polyethylene is a polymer in which a large number of ethylene molecules ($CH_2 = CH_2$) add together to form a molecule

$$R_1-CH_2-CH_2-CH_2 - \cdots - CH_2-CH_2-R_2$$

where R_1 and R_2 are end groups (probably CH_3). Ethylene is the *monomer*.

Let P_n denote a polymer chain with n repeat units that has an active end and hence can still grow (a *live* chain), and let D_n denote a polymer chain with n repeat units that can no longer grow (a *dead* chain). M denotes the monomer. The major chemical steps in free-radical addition polymerization are then propagation: $P_n + M \rightarrow P_{n+1}$ (11.11a)

chain transfer: $P_n + M \rightarrow D_n + P_1$ (11.11b)

termination: $P_n + P_m \rightarrow D_n + D_m$ (11.11c)

There are other steps that are very important with regard to the overall chemistry, but these are sufficient for the purposes of our discussion.

Following the usual practice in this area, we will use the symbols of the chemical species to denote molar concentrations of those species; that is, P_n denotes the concentration of species P_n, etc. Reaction rates are assumed to be proportional to the products of reactant concentrations:

$$r_{p,n} = k_p M P_n \quad (11.12a)$$

$$r_{tr,n} = k_{tr} M P_n \quad (11.12b)$$

$$r_{t,nm} = k_{t,nm} P_n P_m \quad (11.12c)$$

Subscripts p, tr, and t refer to propagation, transfer, and termination, respectively. The rate constants for propagation and transfer are generally taken to be the same for all chain lengths (since the reaction is between the monomer and the "active site" at the chain end), but termination is sometimes taken as depending on the two chain lengths.

The mass balance for species P_n in a well-stirred liquid phase batch reactor, assuming constant volume, is

$$\frac{dP_n}{dt} = k_p M P_{n-1} - k_p M P_n - k_{tr} M P_n - k_{t,n1} P_n P_1$$
$$- k_{t,n2} P_n P_2 - \cdots - k_{t,nN} P_n P_N \quad (11.13)$$

The first term on the right is the rate of formation by addition of monomer to P_{n-1}. The second term is the rate of disappearance by monomer addition to form P_{n+1}, while the third is the disappearance rate through chain transfer to form D_n. The remaining terms represent the rate of disappearance through termination with $P_1, P_2, \ldots, P_N$, respectively, where N is the longest chain in the reactor. Equation (11.13) applies only for $n \geq 2$, and a different equation is required for P_1 to account for the formation of P_1 through chain transfer, Eq. (11.11b). Equation (11.13) can be written more compactly as

$$\frac{dP_n}{dt} = -k_p M (P_n - P_{n-1}) - k_{tr} M P_n - P_n \sum_{m=1}^{N} k_{t,nm} P_m \quad (11.14)$$

As we did in the distillation example, we change nomenclature some-

11.5 AN ALTERNATIVE FORMALISM

what and replace $P_n(t)$ by $P(z,t)$, where $z = v/N$, $0 \le v \le N$. We can identify $P(z,t)$ with P_n when v takes on integral values. Similarly, we replace $k_{t,nm}$ with $k_t(z,\zeta)$. Equation (11.14) is then written at some z as

$$\frac{\partial P}{\partial t} = \frac{-k_p M}{N}\left[\frac{P(z,t) - P(z - 1/N, t)}{1/N}\right] - k_{tr}MP(z,t)$$

$$- NP(z,t)\sum_{n=1}^{N} k_t\left(z,\frac{n}{N}\right) P\left(\frac{n}{N},t\right)\frac{1}{N} \quad (11.15)$$

For large N, the difference quotient in the brackets approximates $\partial P/\partial z$, while the sum approximates the integral, and Eq. (11.15) is approximated by

$$\frac{\partial P}{\partial t} = -\frac{k_p M}{N}\frac{\partial P}{\partial z} - k_{tr}MP - NP\int_{1/N}^{1} k_t(z,\zeta)P(\zeta,t)d\zeta \quad (11.16)$$

Finally, it is more convenient to consider the variables to be functions of v rather than z, noting that $\partial v = N \partial z$, and to write

$$\frac{\partial P}{\partial t} = -k_p M \frac{\partial P}{\partial v} - k_{tr}MP - P\int_{1}^{N} k_t(v,\mu)P(\mu,t)d\mu \quad (11.17)$$

$P(1,t)$ is described by a separate relation and serves as a boundary condition. For large N the upper limit of the integral can be replaced by ∞ with no change in the degree of approximation.

Equation (11.17), and the accompanying equations for other chemical species, can now be solved using any approximation procedure that is appropriate for partial differential-integral equations. Coyle and coworkers have used the finite element method, while Baillagou and Soong have used collocation, to obtain a low-order set of ordinary differential equations. [The equation used by both sets of authors differs somewhat from Eq. (11.17) because of physical or mathematical assumptions.] It is worth noting in passing that Eq. (11.17) can be simplified for the case in which k_t is independent of n and m.

11.5 AN ALTERNATIVE FORMALISM

Most of the derivations of continuous approximations employ a formalism that is somewhat troubling. Consider the term P_{n-1} in Eq. (11.14). If we think of P as a function of n, $P(n)$, then the alternative formalism would expand $P(n - 1)$ in a Taylor series about $P(n)$:

$$P(n - 1) = P(n) + \frac{\partial P}{\partial n} [(n - 1) - n] + \cdots \qquad (11.18)$$

The problem is that we don't know where to stop, because the concept of "higher-order terms" has no meaning; Δn is always equal to unity, as is $(\Delta n)^2$, $(\Delta n)^3$, etc.! Some recent authors writing about batch polymerization have included a second-order term in the expansion, and written

$$P_{n-1} - P_n \approx -\frac{\partial P}{\partial n} + \frac{1}{2} \frac{\partial^2 P}{\partial n^2} \qquad (11.19)$$

This does not seem to be consistent with the order of approximation that is being employed, although Ray claims that the usefulness of the continuous approximation is dependent on the number of terms that is retained. (Baillagou and Soong find that $\partial^2 P/\partial n^2$ is three orders of magnitude smaller than $\partial P/\partial n$ in their calculations, so the problem may be academic.)

BIBLIOGRAPHICAL NOTES

The continuous approximation to a distillation tower grows out of early work on isotope separation. The following two papers contain many references to the earlier literature:

R. F. Jackson and R. L. Pigford, *Ind. Eng. Chem.* **48** (1956) 1020.
H. H. Rosenbrock, *Trans. Inst. Chem. Engrs.* **40** (1962) 376.

Solutions for a variety of cases are contained in

R. L. Pigford, *Dechema Monogr.* **53** (1965) 217.

There are also derivations and some discussion in

L. A. Gould, *Chemical Process Control: Theory and Applications* (Reading, MA: Addison-Wesley, 1969).
H. H. Rosenbrock and C. Storey, *Computational Techniques for Chemical Engineers* (New York, NY: Pergamon Press, 1966).

The equations in Gould and in Rosenbrock and Storey differ in detail from Eq. (11.10).

The continuous approximation for polymerization dates at least to 1954 in

C. H. Bamford and H. Tompa, *Trans. Faraday Soc.* **50** (1954) 1097.

A number of references are contained in

W. H. Ray, *J. Macromol. Sci.–Revs. Macromol. Chem.* **C8** (1972) 1.

More recent work presented at technical meetings by D. J. Coyle, T. J. Tulig, and M. Tirrell and by P. Baillagou and D. S. Soong is unpublished at the time of writing. Katz and coworkers have noted that the continuous approximation for polymerization can be obtained directly from a population balance formulation; see

S. Katz and G. M. Saidel, *AIChE J.* **13** (1967) 319.
H. M. Hulbert and S. Katz, *Chem. Eng. Sci.* **19** (1964) 555.

Katz and Saidel do not include a second derivative term.

12

CASE STUDY: FIBER SPINLINE

12.1 INTRODUCTION

Fiber spinning is a process in which an extruded liquid filament, usually a polymeric liquid, is continuously drawn and simultaneously solidified to form a continuous fiber. The process is described briefly in Section 2.5. There are three fundamental processes for the manufacture of synthetic fibers: melt, wet, and dry spinning. Melt spinning is restricted to polymers that are thermally stable above the melting point, and solidification is effected by cooling to below the crystallization or vitrification temperature. The liquid filament is extruded as a solution in wet and dry spinning, and solidification is effected by solvent removal in a coagulation bath or by evaporation to a hot gas, respectively. We shall restrict our attention here to melt spinning of amorphous (noncrystalline) polymers, which includes the commonly used polyethylene terephthalate (a polyester) but not polypropylene. Glass spinning is also included, with minor modifications as noted in the text.

As noted in Section 2.5, the important physical properties of the solid filament appear to correlate with the optical birefringence. The birefringence is proportional to the stress. (Birefringence is often used as an experimental technique for measuring stress.) Thus, the physical properties of interest correlate with the stress state in the spinline at solidification. One goal of a spinline model, therefore, is to be able to determine the effect of varying process conditions and polymer properties on the stress; the published literature establishes widespread use in the industry of simulations embodying such models.

The second goal of a modeling study of the spinline is to develop an understanding of the transients that occur during spinning. Diameter nonuniformity is a major processing problem. Diameter nonuniformity is most likely caused by propagation through the process of disturbances such as turbulent fluctuations in the cooling air, though instabilities might sometimes occur as well (cf. Fig. 2.8). We shall deal with both phenomena in this chapter.

12.2   THIN FILAMENT EQUATIONS

12.2.1   Problem Definition

The melt spinning process involves simultaneous momentum and heat transfer. As such, it differs qualitatively from most of the other process models discussed here, where explicit assumptions about fluid motion (e.g., the perfect mixing assumption) obviate the need to consider momentum conservation. The most fundamental description of such a process essentially starts with the equations in Table 6.1, with the following (conceptually) minor changes:

- The mass and momentum equations (I and II, respectively) must be written in an axi-symmetric cylindrical (rz) coordinate system.
- The constitutive equation relating the stress to the deformation rate will usually be more complex than the Newtonian fluid (III), and hence the Navier-Stokes equations (IV) will not apply.
- The dissipation term in the energy equation, which is written for a Newtonian fluid, will be different; this is not an important consideration, since dissipation is usually negligible in any event.

A complete fundamental analysis of a spinline, even for a Newtonian fluid such as glass, is a formidable task. The reasons may be deduced from the schematic in Fig. 12.1. The fluid is sheared in the spinneret (the capillary from which it is extruded); because fluid adheres to the stationary capillary wall there is a substantial velocity gradient over the cross section, and a large wall shear stress. As soon as the fluid exits, the wall shear stress is removed discontinuously; in the absence of any significant drag at the outer surface (aerodynamic drag is insufficient) there is no shearing force to sustain a velocity gradient, and the velocity should rearrange to a uniform profile. The velocity rearrangement can cause a *swelling* of the extruded filament, despite an imposed force that draws it down.

Because of the transition from a flow that is mostly shearing to one that is mostly the extension of a uniform filament, analytical solutions of

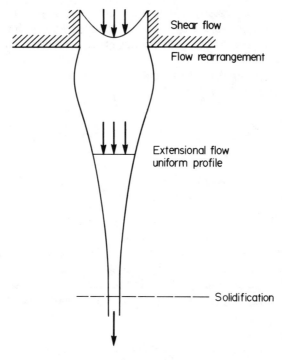

FIG. 12.1.
Schematic of melt spinning.

the type found in fluid mechanics texts are not possible even in the absence of heat transfer. An approximation leading to what is known as the *thin filament equations* does result in a useful process model that is applicable to spinline behavior everywhere except in the immediate region of the spinneret; this excluded region extends only one or two spinneret diameters down from the point of extrusion, while the melt drawing region will typically be hundreds of diameters in length.

12.2.2 Derivation

The thin filament approximation is similar to the lubrication approximation in classical fluid mechanics and the theory of thin plates and shells in classical elasticity. We make use of the fact that, following the initial region of velocity rearrangement, the filament radius $R(z)$ changes slowly with axial distance z, in a sense that must be defined. A number of approaches are

12.2 THIN FILAMENT EQUATIONS

then possible. The most rigorous is to carry out a formal perturbation expansion in a small parameter; this is tedious, and has been done only in some special cases, but it does establish the sense in which dR/dz must be small. Alternatively, it is possible to *average* the momentum, continuity, and energy equations over the filament cross section at each axial position; the averaged axial velocity, for example, is defined

$$v(z) = \frac{1}{\pi R^2} \int_0^R 2\pi r v_z(r,z) dr$$

where v_z is the axial component of the velocity vector, **v**. A number of assumptions are required in this averaging, including the following:

- The normal to the surface differs only slightly from the radial coordinate direction.
- $(d/dz)\int_0^R r^2 \tau_{rz} \, dr$ is small relative to the tensile force in the filament. (τ_{rz} is the shear stress in the z-direction on a surface at constant r.)
- The axial velocity at the outer surface is approximately equal to the average axial velocity.

The averaging is a bit delicate, and we shall not go through the details here. We can obtain a sufficient understanding of the basic spinline model through macroscopic balances on a differential segment of the filament (equivalent to a *free body diagram* in classical mechanics). The control volume is shown in Fig. 12.2. The principle of conservation of mass states that the rate of change of mass in the control volume equals the mass flow rate in at z less the mass flow rate out at $z + \Delta z$. Mass flow rate is $\pi R^2 \rho v$, where ρ is the density and v is the average velocity, while the total mass in the control volume is $\pi R^2 \Delta z \rho$; thus,

$$\frac{\partial}{\partial t} \pi R^2 \Delta z \rho = \pi R^2 \rho v \big|_z - \pi R^2 \rho v \big|_{z+\Delta z}$$

The time derivative is a partial derivative because position z is being held constant. Dividing by Δz and taking the limit as $\Delta z \to 0$ then gives

$$\frac{\partial}{\partial t} R^2 \rho = -\frac{\partial}{\partial z} R^2 \rho v \qquad (12.1)$$

At steady state, Eq. (12.1) simplifies to

$$\text{steady state:} \quad R^2 \rho v = \text{constant} \equiv w \qquad (12.2)$$

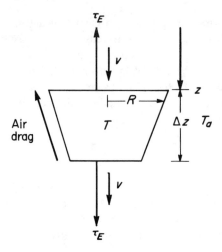

FIG. 12.2.
Control volume for conservation equations.

The principle of conservation of linear momentum states that the rate of charge of momentum in the control volume equals the rate of momentum flow in, less the rate of momentum flow out, plus the sum of the imposed forces. Momentum conservation is a vector relationship, and we are interested only in the z-component. Momentum per unit mass is simply the velocity. Thus, the total z-momentum in the control volume is $\pi R^2 \Delta z \rho v$, while the momentum flow rate is $\pi R^2 \rho v^2$ (mass flow rate $\pi R^2 \rho v$ multiplied by momentum per unit mass). The most important of the forces is $\pi R^2 \tau_E$, where τ_E is the tensile stress exerted on the control volume by the surrounding fluid. The direction shown in Fig. 12.2 represents the usual convention in mechanics, in which a positive stress puts a control volume in tension; thus, the tensile force at z points in the negative z-direction. The other important forces are air drag, which acts on the outside surface opposite to the direction of motion, and possibly gravity. Surface tension forces because of the finite surface curvature are usually unimportant in melt spinning. We therefore have*

$$\frac{\partial}{\partial t} \pi R^2 \Delta z \rho v = \pi R^2 \rho v^2 |_z - \pi R^2 \rho v^2 |_{z+\Delta z} + \pi R^2 \tau_e |_{z+\Delta z}$$

$$- \pi R^2 \tau_E |_z - (\tfrac{1}{2} \rho_a v^2)(2\pi R \, \Delta z) \, C_D + \rho \pi R^2 \Delta z g$$

---

*Because of the changing cross section, there will be a component of atmospheric pressure acting in the axial direction on the projected surface. This is readily shown to be balanced by a pressure contribution to τ_E, and there is no loss in generality in taking the atmospheric pressure to be zero.

12.2 THIN FILAMENT EQUATIONS

Here we have followed the usual convention and written the drag force as the product of an impact pressure term ($\frac{1}{2}\rho_a v^2$, where ρ_a is the density of the air), the surface area over which the force operates ($2\pi R\,\Delta z$), and an aerodynamic *drag coefficient*, C_D. g is the component of the gravitational acceleration in the spinning direction, which we usually take to be vertical. Dividing by Δz and taking the limit as $\Delta z \to 0$ then results in the equation

$$\frac{\partial}{\partial t} R^2\rho v = -\frac{\partial}{\partial z} R^2 \rho v^2 + \frac{\partial}{\partial z} R^2 \tau_E - \rho_a v^2 R C_D + R^2 \rho g \quad (12.3a)$$

This equation is simplified somewhat by use of Eq. (12.1), yielding

$$R^2 \rho \frac{\partial v}{\partial t} = -R^2 \rho v \frac{\partial v}{\partial z} + \frac{\partial}{\partial z} R^2 \tau_E - \rho_a v^2 R C_D + R^2 \rho g \quad (12.3b)$$

At steady state, $\partial v/\partial t = 0$ and $R^2 \rho v = w$, a constant, giving

$$\text{steady state:} \quad -w\frac{dv}{dz} + \frac{d}{dz} R^2 \tau_E - \rho_a v^2 R C_D + R^2 \rho g = 0 \quad (12.4)$$

The equation of conservation of energy states that the rate of change of the total energy contained in the control volume equals the net rate at which energy enters the control volume, plus the rate at which work is done on the control volume by the surroundings. The net rate at which energy enters includes flow in and out, as well as heat transfer to the surroundings. Normally, only internal energy needs to be considered; total internal energy in the control volume is $\underline{U}\pi R^2 \Delta z \rho$, where $\underline{U}$ is internal energy per unit mass. As in Section 6.2, the work to move fluid into and out of the control volume can be combined with the flow of internal energy as flow of *enthalpy*, $\pi R^2 v \underline{H}$, where $\underline{H} = \underline{U} + p/\rho$ is the enthalpy per unit mass.* We than have

$$\frac{\partial}{\partial t} \underline{U}\pi R^2 \,\Delta z \rho = \pi R^2 v \underline{H}\big|_z - \pi R^2 v \underline{H}\big|_{z+\Delta z} - h 2\pi R \,\Delta z (T - T_a)$$

$T(z)$ is the average filament temperature, while T_a is the temperature of the surrounding air; the heat loss to surroundings is written in terms of a local heat transfer coefficient, h. Dividing by Δz and taking the limit $\Delta z \to 0$ leads to

---

*We neglect here the work done on the control volume by viscous stresses, which leads to the viscous dissipation term in the energy balance. Dissipation is almost always negligible in melt spinning.

$$\frac{\partial}{\partial t} \underline{U} R^2 \rho = - \frac{\partial}{\partial z} \underline{H} R^2 v - 2hR(T - T_a)$$

It is now convenient to replace $\underline{U}$ by $\underline{H} - p/\rho$ and to use Eq. (12.1) to obtain

$$R^2 \rho \frac{\partial \underline{H}}{\partial t} - \frac{\partial}{\partial t} pR^2 = - R^2 \rho v \frac{\partial \underline{H}}{\partial z} - 2hR(T - T_a)$$

The term $\partial(pR^2)/\partial t$ can usually be ignored for liquids. If $\underline{H}$ is taken to depend only on temperature then $\partial \underline{H} = c_p \, \partial T$, and we obtain, finally,

$$R^2 \rho c_p \frac{\partial T}{\partial t} = - R^2 \rho v c_p \frac{\partial T}{\partial z} - 2hR(T - T_a) \quad (12.5)$$

At steady state we have

$$\text{steady state:} \quad wc_p \frac{dT}{dz} = - 2\pi hR(T - T_a) \quad (12.6)$$

The equations derived in this way are equivalent to those obtained through the formal averaging process, but the assumptions are less clear.

12.2.3 Stress Constitutive Equation

The total tensile stress, τ_E, is related to the deformation. Stress constitutive equations are written only for the *extra stress,* which is that portion of the stress that excludes the isotropic *pressure.* τ_E is related to the extra stress in the flow direction, τ_{zz}, through the relation $\tau_E = -p + \tau_{zz}$. The discipline of *rheology* is concerned with establishing the relationship between stress and deformation. The equation for a Newtonian fluid is given in Table 6.1, Eqs. (III); this equation assumes that the stress is a linear symmetric function of the velocity gradients, and appears to be applicable to all gases and low molecular weight liquids. The Hookean solid is a solid in which the stress is a linear symmetric function of the strain (i.e., of the displacement gradients), and applicable to many solid materials well below the yield point.

The uniform stretching of a cylinder is a classical problem both in Newtonian fluid mechanics and in Hookean (linear) elasticity. In the former, the total tensile stress equals three times the shear viscosity times the stretch rate; that is, $\tau_E = 3\eta \, \partial v/\partial z$. In the latter, for an incompressible solid (Poisson's ratio = 0.5), the total tensile stress equals three times the shear mod-

12.2 THIN FILAMENT EQUATIONS

ulus (denoted G) times the strain; equivalently, since $\partial v/\partial z$ is the rate of strain (velocity is the time derivative of displacement, so $\partial v/\partial z$ is the derivative of displacement gradient), the expression for the tensile stress in a uniform cylinder can be written $\partial \tau_E/\partial t = 3G\, \partial v/\partial z$.

In the thin filament approximation the stretching process is viewed as being approximately that of stretching a uniform (i.e., constant cross-section) cylinder, so we close the model equations with the following constitutive equation if we take the fluid to be Newtonian:

$$\text{Newtonian fluid: } \tau_E = 3\eta \frac{\partial v}{\partial z} \qquad (12.7)$$

The shear viscosity η will normally be temperature dependent, greatly so for a molten polymer or for an inorganic glass. Most polymeric liquids do not behave like Newtonian liquids, however. For example, Eq. (12.7) indicates that the tensile stress jumps instantaneously following a step change in deformation rate, while it is known that the stress in most polymer melts adjusts over a time scale that can be as long as seconds following a sudden change in deformation rate. The Hookean solid, on the other hand, predicts a jump in stress rate of change following a sudden change in deformation rate, but the stress never reaches a steady value unless the deformation rate becomes zero.

Polymer melts are *viscoelastic*, in that they show deformation behavior that contains features of responses characteristic of both viscous liquids and elastic solids. A useful conceptual model (in the sense of Section 1.2.3) is that of a Newtonian fluid and a Hookean solid responding in series to an imposed stress, as symbolized by the dashpot (viscous) and the spring (elastic), respectively, in Fig. 12.3. The force in the dashpot is $\mu \dot{z}_1$, where μ is proportional to the dashpot fluid viscosity, while the force in the spring is kz_2, where k is the spring constant. But both elements experience the same force, F, so $F = kz_2 = \mu \dot{z}_1$; the first relation can be written equivalently as $\dot{F} = k\dot{z}_2$. Total displacement is $z = z_1 + z_2$, or $\dot{z} = \dot{z}_1 + \dot{z}_2 = F/\mu + \dot{F}/k$; $F + (\mu/k)\dot{F} = \mu\dot{z}$. By analogy, replacing F with τ_E, μ with 3η, k with $3G$, and $\dot{z}$ with $\partial v/\partial z$, we would expect the tensile stress to satisfy the equation

$$\tau_E + \frac{\eta}{G}\dot{\tau}_E = 3\eta \frac{\partial v}{\partial z} \qquad (12.8)$$

Equation (12.8) does roughly represent the response of some polymer melts, but there is an ambiguity that must be resolved in the definition of the time derivative. The Hookean solid (or equivalently, the linear spring) is only a valid constitutive equation for small deformations, while the de-

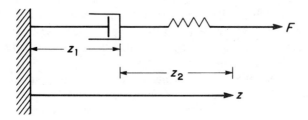

FIG. 12.3.
Spring and dashpot in series.

formations in the processing of polymer melts are large; area reduction ratios of more than 100 are common in spinning, for example. A proper three-dimensional formulation of Eq. (12.8) that is applicable to finite deformations can be constructed. It follows from symmetry and invariance arguments (e.g., "the material response must be independent of the frame of reference of the observer") that the general form which reduces to Eq. (12.8) for an extensional flow is not properly invariant; the properly invariant generalization of Eq. (12.8) for extensional flows, known as the *convected Maxwell model*, is

$$\tau_E + \frac{\eta}{G}\left[\frac{\partial \tau_E}{\partial t} + v\frac{\partial \tau_E}{\partial z} - 2\tau_E\frac{\partial v}{\partial z} - 3p\frac{\partial v}{\partial z}\right] = 3\eta\frac{\partial v}{\partial z} \quad (12.9a)$$

$$p + \frac{\eta}{G}\left[\frac{\partial p}{\partial t} + v\frac{\partial p}{\partial z} + p\frac{\partial v}{\partial z}\right] = -\eta\frac{\partial v}{\partial z} \quad (12.9b)$$

The nonlinear terms vanish for sufficiently small deformations, and Eq. (12.9a) reduces to Eq. (12.8).* The shear modulus G is usually only slightly temperature or deformation dependent, but the viscosity η will depend on T and perhaps on the deformation rate as well. For polyesters used for melt spinning, the viscosity appears (to a good approximation) to depend only on temperature, and not on deformation rate, but such is not the case for polyolefines.

Equation (12.9) has one interesting feature that stems from the fact that it is equivalent to a material having a linear spring in series, which can stretch to any length with sufficient force. Consider steady uniform stretching of a cylinder, where $\partial/\partial t = 0$, $\partial v/\partial z =$ constant, and we expect no

---

*Equation (12.9) is not the most general properly invariant constitutive equation that reduces to Eq. (12.8) for small deformations, but it is a form that follows from molecular considerations and is adequate for our purposes here.

12.2 THIN FILAMENT EQUATIONS

spatial dependence of τ_E or p. In that case, Eq. (12.9) can be solved to yield

$$\tau_E = \frac{3\eta \dfrac{\partial v}{\partial z}}{\left(1 - 2\dfrac{\eta}{G}\dfrac{\partial v}{\partial z}\right)\left(1 + \dfrac{\eta}{G}\dfrac{\partial v}{\partial z}\right)} \quad (12.10)$$

The tensile stress can be very large relative to that in a Newtonian fluid having the same viscosity (Eq. 12.7), and is predicted to become infinite at the finite stretch rate $\partial v/\partial z = G/2\eta$. [Clearly an infinite stress cannot build up in a finite time, so Eq. (12.10) really indicates that no steady stress is possible above a critical stretch rate.]

Rheologists have developed a number of constitutive equations that avoid the "infinity" paradox of the convected Maxwell model. Most start from a structural model of the mechanics of entanglements of macromolecules in a polymer melt, using statistical mechanics to arrive at the final formulation. One such equation, which is used in the simulations shown subsequently in this chapter, is the *Phan-Thien/Tanner model*; when specialized to an extensional flow the equation has the form

$$\tau_E = \sum_{i=1}^{N} \tau_{Ei}, \quad p = \sum_{i=1}^{N} p_i \quad (12.11a)$$

$$\exp\left(\frac{\epsilon[\tau_{Ei} + 3p_i]}{G_i}\right)\tau_{Ei} + \frac{\eta_i}{G_i}\left[\frac{\partial \tau_{Ei}}{\partial t} + v\frac{\partial \tau_{Ei}}{\partial z} - 2\tau_{Ei}\frac{\partial v}{\partial z} - 3p_i\frac{\partial v}{\partial z}\right] = 3\eta_i\frac{\partial v}{\partial z} \quad (12.11b)$$

$$\exp\left(\frac{\epsilon[\tau_{Ei} + 3p_i]}{G_i}\right)p_i + \frac{\eta_i}{G_i}\left[\frac{\partial p_i}{\partial t} + v\frac{\partial p_i}{\partial z} + p_i\frac{\partial v}{\partial z}\right] = -\eta_i\frac{\partial v}{\partial z} \quad (12.11c)$$

The N terms account for the fact that the molecules have a number of independent modes of relaxation. $\{\eta_i\}$ and $\{G_i\}$ are material parameters that can be determined (in principle) from frequency response experiments leading to Bode diagrams like Fig. 1.2. The shear viscosity $\eta = \Sigma\eta_i$; the model equations as written do not include a parameter that accounts for the deformation rate dependence of the viscosity, and has thus already been specialized to polymers like polyesters. ϵ is a new material parameter that prevents the infinite stress; for $N = 1$, $\tau_E \to G\ell n(2\eta\ \partial v/\partial z/G)/\epsilon$ as $\partial v/\partial z \to \infty$.

It is appropriate to say a few words here about rheological measurement. Five or more relaxation modes ($N \geq 5$) are typically required to describe the dynamical behavior of industrial polymers like polyethylene and polystyrene, and measurements of $\{\eta_i\}$ and $\{G_i\}$ for these polymers are reliable. ϵ can be measured only in a stretching experiment, and data are extremely limited. Data on polyethylene terephthalate (PET), for which spinline simulations are described subsequently and compared to pilot plant data, are extremely difficult to obtain because of the low shear viscosity, high shear modulus, and extreme environmental sensitivity. Data are available to *estimate* the best values of $\{\eta_i\}$ and $\{G_i\}$ for $N = 2$, though only the best fit to $N = 1$ can be considered to be reliable. No instrumentation exists that will permit a direct measurement of ϵ. This basic problem in rheological measurement will be an important consideration when we consider model sensitivity.

12.2.4 Phenomenological Coefficients

There are two phenomenological coefficients in the model that describe the interaction between the spinline and the surrounding quench air: C_D, the drag coefficient; and h, the heat transfer coefficient. These coefficients will depend on the local velocity and filament radius, as well as operating conditions. Some insight into the proper functional form can be obtained using boundary layer theory, as in Section 6.6; experimental data are ultimately required for parameter evaluation, and there is considerable scatter in the data. Furthermore, measurements under actual spinning conditions are difficult to make, and indirect experiments presumed to be equivalent (e.g., on moving or stationary solid wires) have often been used.

Drag coefficient data correlate quite well with

$$C_D = \beta \, \text{Re}^{-0.61} \tag{12.12}$$

where $\text{Re} = 2\rho_a v R / \eta_a$ is the Reynolds number based on air density (ρ_a) and viscosity (η_a). The midrange of reported data corresponds to $\beta = 0.6$, but the best value may be closer to 0.37, and perhaps as low as 0.27. Thus, the functional form seems to be established, but there is an uncertainty of about a factor of two in the actual value. Here 0.6 is used as the base case.

The heat transfer coefficient is expected to correlate in the form of the Nussult number as a unique function of the Reynolds number; this expectation is based on the fact that natural convection should not be important at the speeds at which spinning takes place, and the Prandtl number dependence (cf. Eq. 6.43) will be suppressed since heat transfer is always to the same medium. Most experimental correlations of heat transfer from a mov-

12.2 THIN FILAMENT EQUATIONS

ing filament into quiescent air are of the form

$$\frac{hR}{k_a} = \gamma \, \text{Re}^a \tag{12.13}$$

where k_a is the thermal conductivity of the air and a ranges from 0.2 to 0.5. The Colburn or Reynolds analogies between mass, heat, and momentum transport can be applied here to give an exponent $a = 0.39$ and $\gamma = \beta/4$; the exponent is certainly in the correct range, though the coefficient might be low. The best of the published correlations seems to be that of Kase and Matsuo, with $a = \frac{1}{3}$ and $\gamma = 0.21$, but these data were taken for air moving past a stationary cylinder. Kase and Matsuo have included a term to account for the effect of the transverse velocity v_a of the quench air, obtained by doing independent experiments with flow parallel and transverse to the wire, to obtain a final result

$$\frac{hR}{k_a} = 0.21 \, \text{Re}^{1/3} \left[1 + \left(\frac{8v_a}{v}\right)^2 \right]^{1/6} \tag{12.14}$$

Most applications have been proprietary, but there are suggestions in the published literature that this correlation is in essential agreement with industrial data, is certainly adequate for predicting important trends, but that there might be quantitative differences from the best heat-transfer data obtained on pilot-scale spinline experiments. Application to inorganic glass spinning will require modifying the correlation to include the effect of radiative heat transfer.*

12.2.5 Boundary Conditions

Establishment of boundary conditions for these equations is not straightforward; while information is available at the spinneret, the model is valid only downstream of the region of velocity rearrangement. The downstream boundary conditions are straightforward as long as we are dealing only with an amorphous material that solidifies rapidly at a critical temperature T_f (usually close to the glass transition temperature for amorphous polymers). The take-up speed v_f is fixed by the throughput rate and the required filament diameter. If the point of solidification is denoted z_f, then we have

---

*The rate of radiative heat transfer is proportional to $T^4 - T_a^4$. Since this can be factored to $[(T^2 + T_a^2)](T + T_a)(T - T_a)$, and the absolute temperatures T and T_a do not differ significantly, the radiation heat-transfer rate is approximately proportional to $T - T_a$ and hence can be included with the convective transport.

$$z = z_f : T = T_f \quad \text{(12.15a)}$$

$$v = v_f \quad \text{(12.15b)}$$

It is important to note that z_f cannot usually be fixed, but is defined by Eq. (12.15). If the solidification point can be fixed independently—by drawing the molten filament into a cold liquid bath, for example, effecting rapid solidification, or by taking the filament up on a roll while still in the liquid state—then Eq. (12.15a) must be removed.

The fluid emerging from the spinneret is undergoing a transition from confined shear to free-surface uniform extension. This rearrangement region can be analyzed using finite-element methods to solve the full continuity and momentum equations, and these calculations are discussed subsequently. Solutions of the thin filament equations were carried out for years before the finite-element analyses were available, however, and this pattern is the typical one. It is therefore important to deal with the problem directly as it evolved historically.

The origin of the spatial coordinate for the thin filament equations should be at a point where the velocity rearrangement is essentially complete and all variables are uniform over the filament cross section. This would typically be one or two spinneret diameters downstream. The initial temperature is always taken to be the extrusion temperature. This inconsistency should not be important on a commercial spinline in which the melt zone is nearly a meter in length, but it could be quite important on a short spinline used for laboratory experiments. The flow rate will be known, but the area (and hence the velocity) at the origin cannot be specified precisely. Most simulations have used the spinneret area to define the initial area, though one industrial simulation reports using a correlation based on the area change measured during extrusion without axial tension. Thus, we have

$$z = 0: R = R_0, \quad \text{given (usually spinneret radius)} \quad \text{(12.16a)}$$

$$v = v_0, \quad \text{given (usually mean spinneret velocity)} \quad \text{(12.16b)}$$

$$T = T_0 \quad \text{(extrusion temperature)} \quad \text{(12.16c)}$$

These boundary conditions are sufficient for a Newtonian fluid, in which the stress is given by Eq. (12.7). The momentum equation (12.3) or (12.4) is then seen to be second order in v, with v_0 and v_f specified at the two ends, and a physically based iterative computational approach presents itself: specifying the initial velocity derivative is equivalent (in the absence of significant gravitational force) to specifying the tension at the top of the spinline. Either tension or takeup speed can be specified for a fixed flow rate, but not both. Thus, the problem is converted to an initial-value problem by specifying the tension, and the tension is adjusted until the downstream boundary condition is satisfied.

If the fluid is truly viscoelastic, described for example by Eqs. (12.11), then initial conditions must be specified for all τ_{Ei} and p_i. Consider first the case for which all $G_i \to \infty$ (i.e., the viscoelasticity is negligible). Then $p_i/\tau_{Ei} \to -1/3$. For finite G_i, however, p_i decreases rapidly to zero in uniform stretching ($\partial p_i/\partial z = 0$, $\partial v/\partial z =$ constant), while τ_{Ei} increases rapidly; thus, the initial value of p_i is usually taken equal to zero for all i. For $N = 1$ the calculated profiles appear to be insensitive to the initial value of the ratio p_i/τ_{Ei}. For $N > 1$ it is also necessary to specify the distribution τ_{Ei}/τ_{EN} at $z = 0$. In the spinneret this ratio varies from $(\eta_i G_N/\eta_N G_i)^2$ to $(\eta_i/\eta_N)^2(G_N/G_i)$, depending on the shape of the viscoelastic spectrum. We thus complete the description with the stress initial conditions

$$z = 0: \tau_{Ei}/\tau_{EN} = \text{given}\left(\text{between } \frac{\eta_i^2 G_N}{\eta_N^2 G_i} \text{ and } \frac{\eta_i^2 G_N^2}{\eta_N^2 G_i^2}\right);$$

$$i = 1, 2, \ldots, N-1 \quad \textbf{(12.17a)}$$

$$p_i = 0 \quad \textbf{(12.17b)}$$

The data for PET are adequate only to estimate a two-term approximation ($N = 2$) to a *wedge spectrum*, for which $\eta_1 = \eta_2$ and $\tau_{E1}/\tau_{E2} = G_2/G_1$, and this is the initial condition used in the simulations discussed subsequently.

12.3 ASYMPTOTIC SOLUTIONS

12.3.1 Introduction

A number of asymptotic solutions can be obtained for limiting cases of the thin filament equations. These are useful both because of the insight that they provide into the system behavior and as a guide to numerical solutions of the full equations. Determination of possible parameter ranges in which there might be numerical sensitivity is always important; for this equation set there are parameter values for which a solution to the boundary value problem posed here does not exist.

12.3.2 Isothermal, Low-Speed Steady Spinning

At low spinning speeds (in practice, for $v_f < 750$–1000 m/min), the inertial and air drag terms in Eq. (12.4) can be neglected. Since gravity can also be neglected in most cases, we then have $R^2\tau_E$ = constant; i.e., the force (area times tensile stress) is a constant. In isothermal experiments, z_f will be fixed by some mechanical means, such as sudden entry into a chill bath or by a takeup roll. ρ will be constant, so Eq. (12.2) becomes R^2 = constant/v, and the combined continuity and momentum equations can be written

$$\tau_e = \text{constant} \times v \qquad (12.18)$$

For the Newtonian fluid we thus find that dv/dz is proportional to v, and the velocity profile must be an exponential; the boundary conditions $v(0) = v_0$, $v(z_f) = v_f$ then fix the solution as

$$v(z)/v_0 = \exp(z \ln D_R/z_f) \qquad (12.19a)$$

$$R(z)/R_0 = \exp(-z \ln D_R/2z_f) \qquad (12.19b)$$

$$\tau_E(z) = 3\eta \ln D_R v(z)/z_f \qquad (12.19c)$$

The *draw ratio*, D_R, is the area reduction ratio. For constant density this is equivalent to

$$D_R = v_f/v_0 \qquad (12.20)$$

[The definition Eq. (12.20) is commonly used for the draw ratio even when the density varies because of a temperature change, though it is clearly incorrect. It is also used in analyzing dynamical situations in which the takeup velocity is constant though the area might be changing, and in that case it does represent the *average* area reduction ratio.] The force is obtained from $\pi R^2 \tau_E$; with Eq. (12.9) this becomes

$$F = 3\pi R^2 \eta \frac{dv}{dz} = \frac{3\eta w}{\rho z_f} \ln D_R \qquad (12.21)$$

Similar solutions exist for simple non-Newtonian fluids that are not viscoelastic, and closed-form solutions exist for the Newtonian fluid with any one of surface tension, gravity, or inertia included. An exact, closed-form solution cannot be obtained for the Maxwell fluid, Eqs. (12.9), even in this limiting case, but perturbation solutions can be obtained. An appropriate perturbation variable is $F/\pi R_0^2 G$; this is the ratio of the imposed tensile stress to the shear modulus, which is a characteristic material stress. The limit $F/\pi R_0^2 G \to 0$ leads to a singular perturbation problem that can be solved to give a correction to Eqs. (12.19) and (12.21). The more interesting case is the limit $F/\pi R_0^2 G \to \infty$; this leads to a regular perturbation which

12.3 ASYMPTOTIC SOLUTIONS

to first order is

$$\frac{v(z)}{v_0} = 1 + \frac{Gz}{\eta v_0}$$
$$+ \frac{3\pi}{2} \frac{R_0^2 G}{F} \left\{ \ln\left(1 + \frac{Gz}{\eta v_0}\right) + \frac{1}{2}\left[1 - \frac{1}{(1 + Gz/\eta v_0)^2}\right] \right\} + \cdots \quad (12.22)$$

$$F = \frac{3\eta w}{4\rho} \left[\frac{2 \ln D_R + 1 - D_R^{-2}}{z_f - \frac{\eta v_0}{G}(D_R - 1)} \right] \quad (12.23)$$

Note that the force is predicted to become infinite at a finite draw ratio equal to $1 + Gz_f/\eta v_0$, and that the velocity becomes linear in this limit. Solutions do not exist for larger values of the draw ratio. This infinite force corresponds to the "linear spring infinity" described in the context of Eq. (12.10), and does not exist when the Phan-Thien/Tanner fluid is used to describe the rheology, nor does it occur even for the Maxwell fluid when inertia is important. To the extent that the Maxwell fluid is a reasonable description of the material, however, it suggests that very large stresses are to be expected relative to a Newtonian fluid of the same shear viscosity, and a far more linear profile. The same conclusions follow when the viscosity in Eq. (12.9) is taken to be deformation dependent, though the form of the asymptotic solution differs somewhat.

There are laboratory experimental data that support the predictions of this asymptotic analysis. Filaments drawn from fluids with identical constant shear viscosities, one a corn syrup and one the same corn syrup with small amounts of a water-soluble polymer added, give stress levels that differ by more than two orders of magnitude under identical spinning conditions. Studies on the spinning of polyethylene and polystyrene melts, as well as polymer solutions, show a far more linear velocity profile than would be obtained for a nonviscoelastic liquid; the data on polyethylene, among the first reported in the literature and predating the analysis leading to Eq. (12.22), are perfectly linear within experimental scatter.

It is possible to use the asymptotic solutions as a means of estimating conditions under which the fluid viscoelasticity will be important. This will occur when the asymptotes intersect and the force given by Eq. (12.21) is determined by the limiting draw ratio computed from Eq. (12.22); the result can be expressed as

$$\frac{F}{3\pi\eta R_0 v_0} \sim \frac{R_0}{z_f} \ln\left(1 + \frac{z_f}{\lambda v_0}\right) \quad (12.24)$$

12.3.3 Temperature Profile

The temperature profile, and hence the length of the liquid spinline to the solidification point, can be estimated quite nicely from the thin filament equations without solution of the full combined kinematics-heat transfer problem. This is because the energy equation is only weakly dependent on the details of the kinematics and hence on the rheology. Equations (12.6) and (12.13) combine to give

$$\frac{1}{T-T_a}\frac{dT}{dz} = -0.42 \frac{k_a}{\rho c_p}\left(\frac{2\rho_a}{\eta_a}\right)^{1/3}\left(\frac{w}{\pi\rho}\right) v(z)^{1/6} \qquad (12.25)$$

Here we have taken the Kase and Matsuo equation for h, with heat transfer into quiescent air. $v^{1/6}$ will vary over the spinline length by no more than 2 or 3. We can therefore obtain bounds on z_f by assuming two limiting cases: $v(z)$ equals v_0 until the end of the spinline and then jumps instantaneously to v_f, and v jumps instantaneously to v_f at $z = 0$. In each case the differential equation can be solved immediately to give

$$1 \geq \frac{z_f [0.42(k_a/\rho c_p)(2\rho_a/\eta_a)^{1/3}(w/\pi\rho)^{-5/6} v_0^{1/6}]}{\ell n\,[(T_0 - T_a)/(T_f - T_a)]} \geq D_R^{-1/6} \qquad (12.26)$$

We see, for example, that the distance to solidification is strongly dependent on throughput but only weakly dependent on takeup speed; this result is in agreement with observations made in practice. The lower bound should be a somewhat better estimate of the actual length, since the spinline will be drawn down rapidly at the maximum temperature (lowest viscosity), and hence most of the heat transfer will take place from the thinner filament. Specific estimates can be obtained by taking more realistic velocity profiles; a linear profile leads to an estimate of the solidification length that is 16% greater than the lower bound (for large D_R), for example. The midpoint of the two bounds in Eq. (12.26) provides a useful means of obtaining a first estimate of z_f if it is needed for numerical computation.

12.4 FINITE ELEMENT SOLUTION

Though our focus in this chapter is on the use of the thin filament equations for developing a process model, it is instructive to record some numerical results using a finite-element method to solve the slow-speed, isothermal flow problem. The finite-element grid is shown in Fig. 12.4. It is not prac-

12.4 FINITE ELEMENT SOLUTION

tical to calculate more than five or ten capillary diameters downstream of the spinneret. The flow at the upstream boundary in the capillary is taken to be fully developed (parabolic velocity profile), while the downstream condition is a specified force. The position of the free surface is found by iteration. The full steady-state microscope momentum, continuity, and stress equations (e.g. those in Table 6.1) are solved, except that the inertial terms are not included in the momentum equation and axisymmetry is assumed.

The computed average velocity for a Newtonian fluid is shown in Fig. 12.5 for a range of imposed forces. The asymptotic exponential solution obtained from the thin filament equations is achieved within one capillary diameter in all cases. It is clear that use of the spinneret velocity as an initial condition would shift the profile by a small amount over the entire length; this would not be important in a simulation of a pilot or commercial scale spinline, where the total distance is hundreds or thousands of capillary diameters, but it could be a factor on a short experimental spinline. All cases extrapolate approximately at $z = 0$ to the velocity corresponding to the minimum in an undrawn jet (*free jet swell*), so a free-swell correlation could be used to provide the initial condition in this case, but this is unlikely to be a general result.

Calculations for the viscoelastic liquids are difficult, and convergent results cannot be obtained for flows with stress singularities when $\eta v_0/R_0 G$ is substantially greater than unity. The reason for this limitation is not fully understood at the time of writing, and the area remains an active one for research in numerical analysis. Results for a Maxwell fluid with $\eta v_0/R_0 G = 1/4$ show that the velocity becomes uniform to within 5% at $z = 1.8 R_0$, which is an only slightly longer rearrangement length than for the Newtonian fluid, but the stresses do not become uniform to within 10% over the cross section until $z = 3.6 R_0$. The computed ratio of $-p/\tau_E$ approaches zero at both of these axial positions for $F/\pi R_0^2 G$ of order unity, indicating that Eq. (12.17b) is a satisfactory boundary condition for large values of $F/\pi R_0^2 G$; the calculations cannot be carried to sufficiently high values to test the approach to the asymptotic solution, Eq. (12.22), but the trend does suggest that a linear profile will develop. The computed draw ratio at $z_f = 10R_0$ is shown in Fig. 12.6 as a function of dimensionless force for the Newtonian

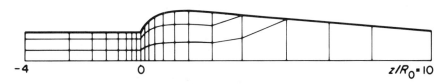

FIG. 12.4.
Finite element grid.

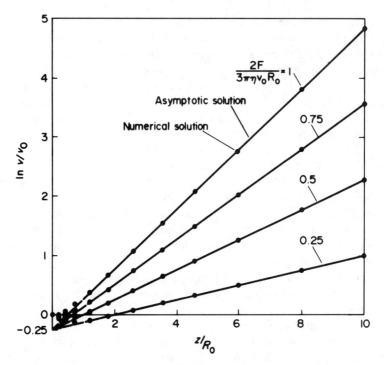

FIG. 12.5.
Computed average velocity, Newtonian fluid (Kennings et al., 1983, copyright American Chemical Society, reproduced with permission).

and Maxwell fluids. The approach to the limiting draw ratio predicted by Eqs. (12.22) and (12.23) is apparent, as is the deviation from Newtonian behavior at a force in the neighborhood of the value given by Eq. (12.24). Similar behavior is observed with more general constitutive equations that permit computation to higher values of the dimensionless group $\eta v_0/GR_0$.

The general conclusion to be drawn from the finite-element calculations is that the thin filament equations, with the boundary conditions developed in Section 12.2.5, should be adequate to describe most steady-state spinline behavior, except perhaps in situations where the spinline length is no more than 10 or so spinneret radii.

12.5 STEADY-STATE SPINLINE SIMULATION

12.5.1 Pilot Plant Results

We describe here an application of the thin filament model to the simulation of a set of pilot plant experiments on a textile-fiber grade polyethylene tere-

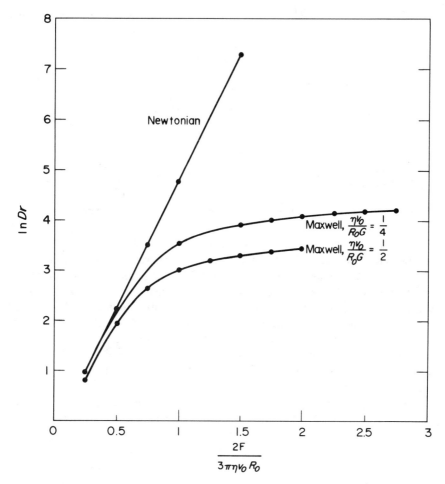

FIG. 12.6.
Draw ratio as a function of dimensionless force, $z_f = 10R_0$ (after Kennings et al., 1983, copyright American Chemical Society, reproduced with permission).

phthalate. The experimental conditions are given in Table 12.1. The PET was taken to have the rheology of a Phan-Thien/Tanner fluid, with ϵ set arbitrarily to 0.015 (the value typically used for polyethylene) in the absence of any data. Literature data on viscosity and mean relaxation time were used to estimate the other parameters in the rheological model for $N = 1$ and 2, assuming a "wedge" spectrum for the latter: $\eta_1 = \eta_2 = \eta/2$; G_1 was set to $5G_2$. The data extend only to 265° C, so extrapolation was required to 70° C, the estimated solidification temperature. A correlation for free jet swell was used to facilitate comparison with another simulation of the same data, so the initial area was set to twice the spinneret area.

TABLE 12.1.
Conditions for spinning experiments.

PET intrinsic viscosity	0.67
Extrusion temperature	295° C
Estimated solidification temperature	70° C
Air temperature	30° C
Air velocity	0.2 m/s
Spinneret hole diameter	0.25 mm
Throughput	1.75×10^{-5} kg/s
Mean extrusion velocity	18.2 m/min

The computed velocity profile is shown in Fig. 12.7 for takeup speeds of 1000 and 3000 m/min, together with the experimental measurements. The agreement is reasonably good, given that the model is based entirely on literature data for the rheology and the heat transfer and drag coefficients, and particularly given the uncertainty in all these measurements. The sharp corner at the point of solidification is a consequence of extrapolating the viscosity to the point of solidification without accounting for the sharp rise near that point; it is a "cosmetic" effect that is easily corrected by using standard viscosity-temperature relations near the neighborhood of the glass-transition temperature (e.g., the *W-L-F equation*). The fit for $N = 2$ is a bit better than for $N = 1$, and it is likely that a more complete representation of the relaxation spectrum would lead to a still better fit. To put the calculation into perspective, however, it should be recalled that the solidification point is relatively insensitive to kinematics for a given throughput (this can be seen in both the data and calculations over the threefold charge in takeup speed), and hence the two endpoints of the velocity function are essentially fixed. The curvature of the velocity-distance curve is therefore a critical comparison for model validation.

12.5.2 Parametric Sensitivity

The sensitivity to the rheology and transport properties can now be analyzed. PET has an essentially constant (deformation-free) viscosity, and it is often treated as a Newtonian fluid. Figure 12.8 shows a simulation of the same experiments using the Newtonian fluid model. Drawdown is much too rapid at the higher speed. The dashed line is a repetition of the same calculation, but with the heat transfer coefficient reduced by 25%, which would be within the margin of uncertainty of most heat transfer correlations. The two cal-

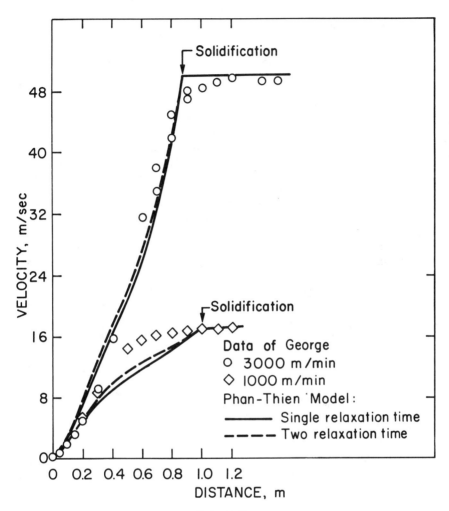

FIG. 12.7.
Computed velocity profiles and spinning data of George (Gagon and Denn, 1981, copyright Society of Plastics Engineers, reproduced with permission).

culations, *viscoelastic fluid* using the Kase-Matsuo heat transfer correlation and *Newtonian fluid* with the reduced heat transfer coefficient, are shown together in Fig. 12.9. On the basis of this calculation alone, given the uncertainties in both the heat transfer and rheological data, it would not be possible to choose between the two simulations. The independent knowledge that PET is slightly viscoelastic seems to make the viscoelastic fluid rep-

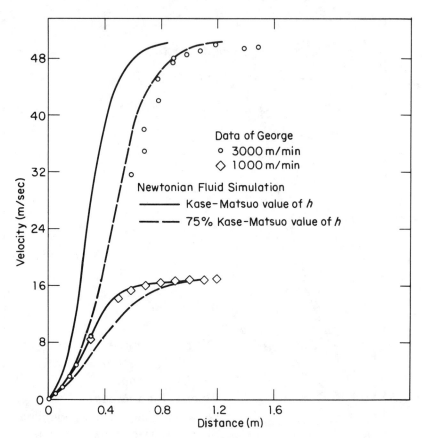

FIG. 12.8.
Newtonian fluid simulation and spinning data of George (Gagon and Denn, 1981, copyright Society of Plastics Engineers, reproduced with permission).

resentation preferable for extrapolation outside the range of these experiments. The issue is a rather deep one, in fact, and requires considerably more discussion, which we defer to the next chapter.

It is useful to examine the sensitivity to the other parameters briefly. $\beta = 0.37$, which was obtained in recent experiments, might be a better estimate of the parameter in the drag coefficient correlation, Eq. (12.12), than the mid-range value of 0.6 used as the base case here. The calculations from Fig. 12.7 are repeated in Fig. 12.10 to show the sensitivity to drag coefficient. There is a small effect, and the lower value is perhaps in slightly better agreement with the experimental data, but the sensitivity is much less than that to the heat transfer coefficient.

12.5 STEADY-STATE SPINLINE SIMULATION

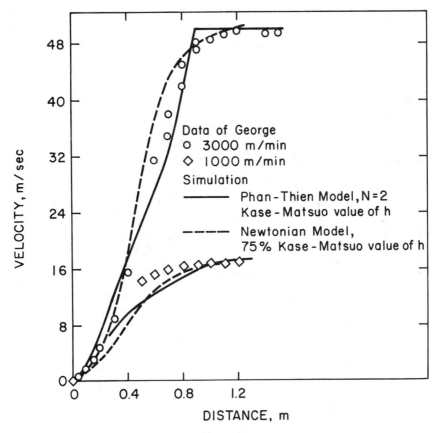

FIG. 12.9.
Comparison of viscoelastic and Newtonian fluid simulations (Gagon and Denn, 1981, copyright Society of Plastics Engineers, reproduced with permission).

The only rheological parameter that cannot be measured directly for PET in the Phan-Thien/Tanner equation is ϵ. Sensitivity to ϵ is shown in Fig. 12.11. There is little difference between $\epsilon = 0$ and 0.015 for this experimental situation, though there can be a pronounced effect under other conditions.

Viscosity measurement for PET is fairly reliable, but measurement of the modulus is far more difficult. The velocity profile and computed stress at solidification (proportional to optical birefringence) are shown for $N = 1$ in Figs. 12.12 and 12.13, respectively, as functions of G_b/G, where G_b is the "base-case" value. A small amount of viscoelasticity change goes a long way here, in that the calculations are insensitive to the value of G over a

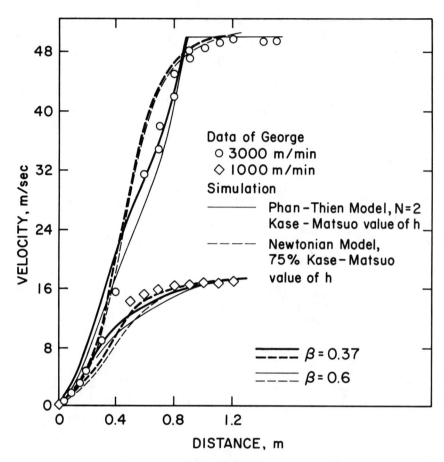

FIG. 12.10.
Sensitivity of spinline model to air drag (Gagon and Denn, 1981, copyright Society of Plastics Engineers, reproduced with permission).

tenfold range beginning at about 10% of the literature value. This is because a small amount of viscoelasticity is sufficient under these spinning conditions to make the profile approximately linear; because of the fixed endpoints there is little that additional viscoelasticity can do to the profile. It is interesting to note that the computed stress (for fixed heat transfer parameters) does not differ substantially for Newtonian and viscoelastic fluids, but that the stress in the viscoelastic liquid is always *lower* than in the Newtonian liquid of the same viscosity.

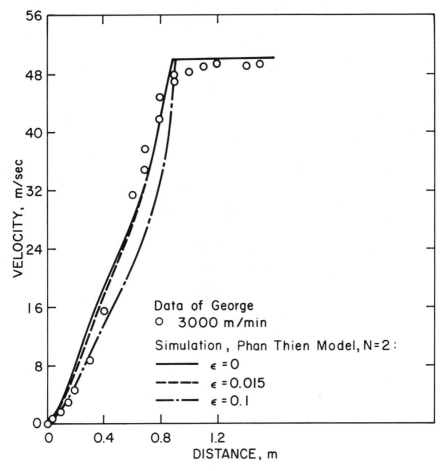

FIG. 12.11.
Sensitivity of spinline model to extensional flow parameter ϵ (Gagon and Denn, 1981, copyright Society of Plastics Engineers, reproduced with permission).

12.5.3 Possible Plant Application

The successful application of the model to the pilot scale experiments encourages application to the analysis of plant problems and possible new designs. The most likely use of such a model would be as a guide to experiments and as diagnostic tool. (The stress at the spinneret, for example, should not be too high, which could cause breakage, or too low, which could

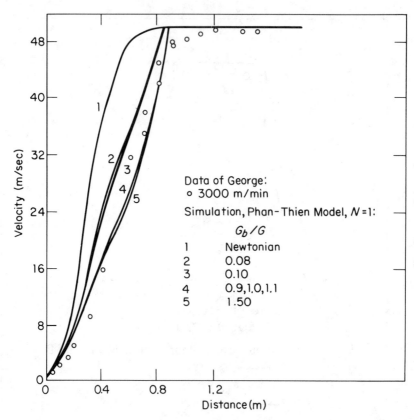

FIG. 12.12.
Sensitivity of spinline model to modulus (Gagon and Denn, 1981, copyright Society of Plastics Engineers, reproduced with permission).

lead to dripping. This stress could be adjusted for new spinning conditions by control of the extrudate viscosity through the temperature.) Little has been published in this area, but it is known that such applications have been made. The sensitivity to details of heat transfer make this a particularly attractive area of application, since quench air flows are usually poorly controlled.

12.5.4 Some Numerical Experiments

One of the uses of a model is to explore new conditions, and to seek out situations where unexpected results are predicted. Such unexpected results

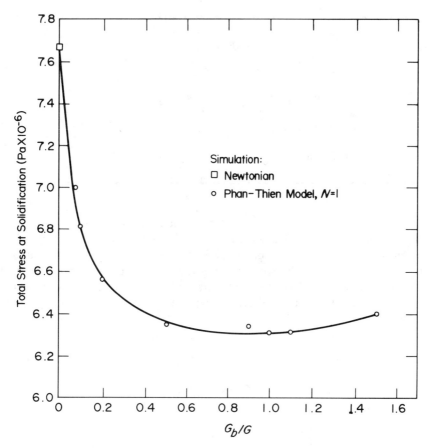

FIG. 12.13.
Sensitivity of stress at solidification to modulus (Gagon and Denn, 1981, copyright Society of Plastics Engineers, reproduced with permission).

might be indicative of new phenomena, or they might simply reflect the limitations of the model. In either case, they represent useful information.

Table 12.2 lists spinning conditions for PET that differ considerably from those in Table 12.1. Four computed stress development profiles for these conditions are shown in Fig. 12.14 for a 0.032-mm diameter (10-*denier*) fiber taken up at 1000 m/min: Newtonian and viscoelastic, both including (*high speed*) and not including (*low speed*) inertia and air drag. The viscoelastic calculations are for $N = 1$ with both the Phan-Thien/Tanner and Maxwell fluid models, which give the same results under these conditions. Inertia and air drag increase the stress in the Newtonian fluid, as would be expected. For the viscoelastic liquid, however, the effect of inertia

TABLE 12.2.
Conditions for numerical experiments.

PET intrinsic viscosity	0.67
Extrusion temperature	280° C
Estimated solidification temperature	60° C
Air temperature	35° C
Air velocity	0.3 m/s
Initial diameter	0.5 mm

and air drag is not only much greater, but it is in the opposite direction! This result is completely counterintuitive. (Indeed, a careful check of the computer code was the first consequence of the calculation.) The calculation is in fact correct, if surprising; air drag retards the velocity profile development, and hence reduces the stretch rate sufficiently to cause a significant decrease in τ_E that more than offsets the stress increase from air drag and inertia.

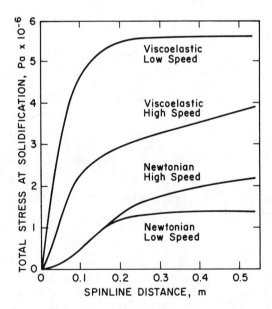

FIG. 12.14.
Computed stress profile, 0.032 mm diameter fiber, 100 m/min, conditions in Table 12.2 (Gagon and Denn, 1981, copyright Society of Plastics Engineers, reproduced with permission).

12.5 STEADY-STATE SPINLINE SIMULATION

The possible effect of an apparently small charge in rheology is shown in Fig. 12.15, where the stress at solidification is plotted versus draw ratio for three rheological models at two takeup speeds, 500 and 4000 m/min,

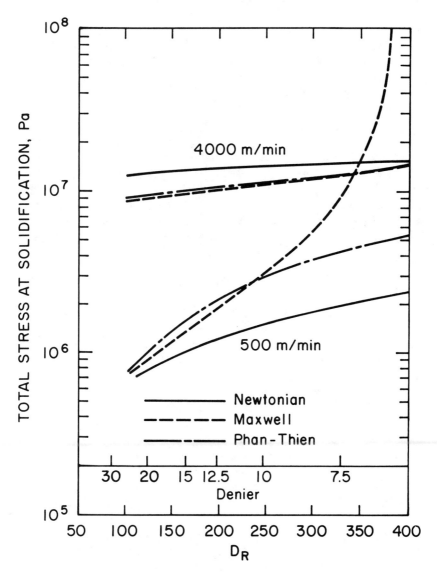

FIG. 12.15.
Computed stress at solidification, conditions in Table 12.2 (Gagon and Denn, 1981, copyright Society of Plastics Engineers, reproduced with permission).

using the spinning conditions in Table 12.2. (Clearly, the extrusion velocity changes with draw ratio at constant takeup speed). Inertia and air drag are negligible at 500 m/min, and the familiar stress infinity is seen at a critical draw ratio for the Maxwell fluid. At high takeup speeds, on the other hand, the stress growth is sufficiently retarded that the stresses are always bounded and are in fact slightly lower than those computed for the Newtonian model. The unexpected interaction between high speed effects and viscoelasticity can lead to some seemingly paradoxical behavior. The Maxwell model predicts that a 7.5-denier fiber can be spun from a 0.5-mm spinneret at speeds above about 3000 m/min and below about 500 m/min, but that stresses in the intermediate range would be excessive. Thus, by lowering throughput and takeup speed, conditions would be reached where the stress would *increase* upon further reduction. This result is nonphysical, and demonstrates the importance of using a stress constitutive equation like Phan-Thien/Tanner which does not allow infinite stresses to develop. (It also demonstrates the need to obtain a good value of ϵ for PET.) The set of numerical experiments demonstrates that viscoelasticity will likely have the greatest effect on the stress at solidification at low and intermediate spinning speeds, rather than at high speeds as might have been expected.

12.6 DRAW RESONANCE

We saw a remarkable set of data on spinning in Fig. 2.8; PET spun through a short air gap into a water chill bath with a constant speed of extrusion and a constant windup velocity emerged as a solid filament with large amplitude variations having a regular periodic structure. This instability, known as *draw resonance,* occurs regularly in isothermal laboratory experiments at a critical draw ratio in the neighborhood of 20. It is less common in commercial spinning, where nonisothermal effects are important, but it is a major limitation in the kinematically similar process of extrusion coating, where even small variations in thickness can have a negative effect on optical properties. We shall touch briefly on draw resonance here because of the belief that the dynamical response of a model, and particularly the transition to an unstable mode, can be a far more sensitive probe of model structure than steady-state response.

The simplest case to analyze is that of isothermal, low-speed spinning of a Newtonian fluid. The time derivative term in Eq. (12.3) is an inertial term, so the transient momentum equation is simply

$$\frac{\partial}{\partial z} R^2 \tau_E = 0 \qquad (12.27)$$

12.6 DRAW RESONANCE

while the continuity equation is

$$\frac{\partial R^2}{\partial t} = -\frac{\partial}{\partial z} R^2 v \qquad (12.28)$$

The stress equation is Eq. (12.7), $\tau_E = 3\eta\, \partial v/\partial z$. Steady-state solutions are given by Eq. (12.19).

We now follow the procedure outlined in Chapter 8. We introduce new variables representing the perturbation from the steady state, as follows:

$$v(z,t) = v_0[\exp(z\,\ell n\, D_R/z_f) + \Psi(z,t)] \qquad (12.29c)$$

$$R^2(z,t) = R_0^2[\exp(-z\,\ell n\, D_R/z_f) + \Phi(z,t)] \qquad (12.29b)$$

$$\tau_E(z,t) = \frac{3\eta v_0}{z_f}[\ell n\, D_R \exp(z\,\ell n\, D_R/z_f) + \Pi(z,t)] \qquad (12.29c)$$

The coefficients that have been factored out simplify the scaling. It is convenient to work with R^2 as the variable, since this is the way in which R always appears in the governing equations. When Eqs. (12.29) are substituted into Eqs. (12.17), (12.28), and (12.7), and nonlinear terms in the perturbations are neglected, the following linear partial differential equations are obtained:

$$\frac{\partial \Psi}{\partial \xi} = \Pi \qquad (12.30a)$$

$$\frac{\partial \Pi}{\partial \xi} = (\ell n\, D_R)\Pi - \ell n\, D_R \exp(2\xi\,\ell n\, D_R)\left[\frac{\partial \Phi}{\partial \xi} - (\ell n\, D_R)\Phi\right] \qquad (12.30b)$$

$$\frac{\partial \Phi}{\partial \xi} = -\exp(-\xi\,\ell n\, D_R)\frac{\partial \Phi}{\partial \theta} - \ell n\, D_R \Phi$$

$$+ \exp(-2\xi\,\ell n\, D_R)[\Pi + (\ell n\, D_R)\Psi] \qquad (12.30c)$$

Here the independent variables are defined as

$$\xi = z/z_f, \qquad \theta = v_0 t/z_f \qquad (12.31)$$

The velocity is specified at $\xi = 0$ and $\xi = 1$, so the perturbation at these points must be zero. Similarly, the initial area is fixed, so the area perturbation must vanish at $\xi = 0$. Thus, the boundary conditions are

$$\Psi(0,t) = \Psi(1,t) = \Phi(0,t) = 0 \qquad (12.32)$$

Equations (12.30) are a homogeneous system with homogeneous boundary conditions, which suggests application of the method of separation of variables. The time dependence is clearly exponential, since time enters only in the form of a derivative. We therefore seek solutions in the form

$$\Psi = \sum_n \{e^{\lambda_n \theta} \Psi_n(z) + e^{\lambda_n^* \theta} \Psi_n^*(z)\} \quad (12.33a)$$

$$\Phi = \sum_n \{e^{\lambda_n \theta} \phi_n(z) + e^{\lambda_n^* \theta} \phi_n^*(z)\} \quad (12.33b)$$

$$\Pi = \sum_n \{e^{\lambda_n \theta} \varpi_n(z) + e^{\lambda_n^* \gamma} \varpi_n^*(z)\} \quad (12.33c)$$

The superscript * denotes the complex conjugate. The modes separate because of the linearity of the equations, and we obtain the following linear ordinary differential equation for each n:

$$\frac{d\psi}{d\xi} = \varpi \quad (12.34a)$$

$$\frac{d\varpi}{d\xi} = (\ell n\, D_R)\tilde{\omega} - \ell n\, D_R(2\xi\, \ell n\, D_R)\left[\frac{d\phi}{d\xi} - (\ell n\, D_R)\xi\right] \quad (12.34b)$$

$$\frac{d\phi}{d\xi} = -[\exp(-\xi\, \ell n\, D_R)\lambda + \ell n\, D_R]\phi$$

$$+ \exp(-\xi\, \ell n\, D_R)[\varpi + (\ell n\, D_R)\psi] \quad (12.34c)$$

$$\psi(0) = \psi(1) = \phi(0) = 0 \quad (12.35)$$

This system of equations always admits the trivial solution $\psi = \tilde{\omega} = \phi = 0$. We are interested only in nontrivial solutions, which will exist for certain eigenvalues λ. The spinning conditions are stable to infinitesimal perturbations if all eigenvalues have negative real parts, since the disturbances will then decay to zero. The spinning conditions are unstable if any one eigenvalue has a positive real part. In principle, this means that all eigenvalues in the infinite set must be examined, since the system is not self-adjoint and the eigenvalues cannot be ordered. In practice, the stability seems to be governed by the eigenvalue with the smallest modulus.

A variety of solution methods can be applied to solve the eigenvalue problem, including reduction to a quadrature. The most straightforward is to solve the problem numerically as an initial value problem and to adjust λ until the downstream boundary condition is satisfied. This can be done interactively, and convergence is rapid. The iteration is aided by the observation that the real part of the first eigenvalue is nearly linear in D_R. It is important to recall that all functions are complex. The initial condition for ϖ can be set as Re $(\varpi(0)) = 1$, Im $(\varpi(0)) = 0$ without any loss of generality, since the equations are homogeneous and the phase angle of $\varpi(0)$ simply sets the phase of the solution. The first two eigenvalues, computed in this way with a Runge-Kutta method, are shown in Table 12.3. The low speed, isothermal spinning process is predicted to become unstable at a draw ratio of approximately 20 ($D_R = 20.218$ is a precise value). The only experiments on a truly Newtonian fluid are confounded by gravity and surface tension, so the predicted critical draw ratio is somewhat higher, but this analysis seems to be essentially in accord with experiment both on the Newtonian fluid and on some very slightly viscoelastic liquids.

The analysis can be repeated with the energy equation included, with both Newtonian and viscoelastic constitutive equations. Cooling increases the critical draw ratio. Viscoelasticity is stabilizing, though there is little effect until a critical value of $\eta v_0 D_R / Gz_f$ is reached. A viscosity that decreases with increasing deformation rate lowers the critical draw ratio. Finally, a nonlinear analysis based on Galerkin's method (Section 10.5) that assumes that the spatial dependence of solutions to Eqs. (12.27) and (12.28) is given approximately by the first eigenfunctions to Eqs. (12.34), but with time-dependent coefficients that reduce to the exponential only in the limit of small perturbations, predicts cyclic behavior like draw resonance above the critical draw ratio, as does direct numerical solution of Eqs. (12.27) and (12.28).

The theoretical results are in general agreement with experiment, lending considerable support to the spinline analysis, though unresolved problems do exist, including deviations from the model predictions that may

TABLE 12.3.
Eigenvalues for low-speed, isothermal, Newtonian spinning.

$\ell n\, D_R$	D_R	λ_1		λ_2	
		Real	Imaginary	Real	Imaginary
1.0	2.718	−3.810	7.671	−6.395	18.332
2.0	7.389	−2.025	10.845	−4.310	26.075
2.95	19.105	−0.120	13.814	−2.347	33.907
3.006	20.210	0.0	13.989	−2.219	34.380
3.15	23.336	+0.309	14.437	−1.884	35.596

be associated with the spinneret region. Figure 12.16 is a plot of diameter variation as a function of spinline length prior to a water bath for a series of experiments on two PET resins at a draw ratio of about 50; the data in Fig. 2.8 are included. Heat transfer is a definite factor at the longer spinning lengths, but the shorter spinlines are nearly isothermal. The solid line is a plot of the growth rate (the real part of λ) as a function of D_R for the given spinning conditions, including heat transfer and taking the PET rheology as Maxwellian. The growth rate becomes negative, indicating stability, at short and long spinning lengths, in accordance with the data, though the stabilization by heat transfer seems to occur at somewhat shorter lengths than predicted by the model. The scale has been chosen so that the maximum passes through the maximum in the data, but this is simply suggestive as there is no reason to expect the most rapid linear growth to correspond to the largest amplitude for the finite disturbance. The dashed line is the approximate nonlinear isothermal theory, and it clearly follows the trend of the data as far as it is applicable.

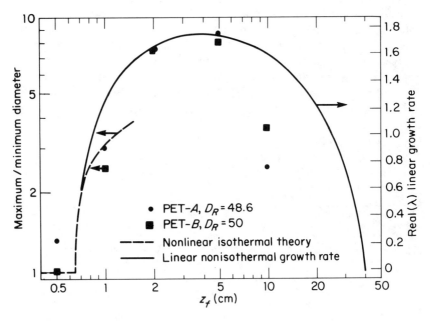

FIG. 12.16.
Result of stability theory and draw resonance data of Ishihara and Kase (Fisher and Denn, 1977, copyright American Institute of Chemical Engineers, reproduced with permission).

12.7 DYNAMIC SENSITIVITY

The analysis of draw resonance represents an interesting intellectual challenge, and it is a good test of the adequacy of a spinline model, but it is of limited industrial significance. Spinline *sensitivity*—the propagation of disturbances, such as quench air turbulence—is a far greater problem with regard to filament uniformity. The traditional means of studying dynamic sensitivity is through a frequency response analysis, where the linearized model equations are subjected to sinusoidal forcing. Because of the linearity, the

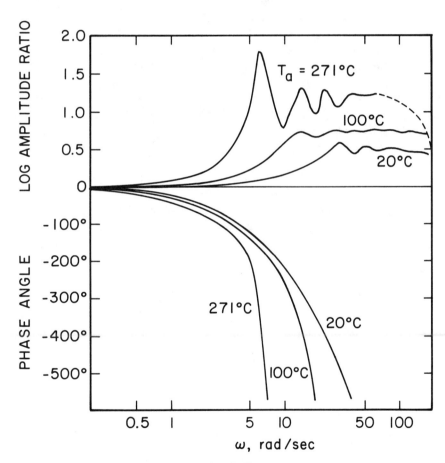

FIG. 12.17.
Bode diagram between takeup area and initial perturbation, Newtonian, slow speed (Kase and Denn, 1978).

response to a sum of sinusoidal forcing functions of different frequencies is the sum of the responses; a weighted sum over all frequencies provides the Fourier sum or integral corresponding to the response to a disturbance having a particular Fourier spectrum. Results are usually presented in the form of a Bode diagram like Fig. 1.2.

Frequency response analyses have been reported thus far only for low-speed spinning of a Newtonian fluid, but including heat transfer. The linear partial differential equations (12.30) must therefore be augmented by the linearized energy equation, and additional terms corresponding to temperature variations must be included (e.g., the variation of viscosity in the stress equation because of a temperature variation). The only conceptual difficulty here is that the solidification length will itself change because of transients in the temperature profile. The linearity of the equations can be exploited, however, to express the downstream boundary condition in terms of a relationship between the velocity and temperature perturbations at the original solidification point so that the equations are solved over a fixed domain. Both direct numerical solution of the partial differential equations and Laplace transformation to a set of ordinary differential equations in the frequency domain have been employed; the former requires a convolution of the time response to step forcing in order to obtain the frequency response.

A typical Bode diagram showing the amplitude ratio and phase shift between takeup area and initial area perturbations is shown in Fig. 12.17 for PET (treated as a Newtonian fluid) with different quench temperatures. The attenuating effect of cooling is clearly seen. The resonant peak will grow in amplitude as a point of instability is approached, but it is important to note that there can be significant amplification of disturbances (large amplitude ratio) even on a spinline (or any process!) that is stable.

12.8 CONCLUDING REMARKS

The fiber spinline illustrates nicely the way in which material constitutive properties and transport coefficients dominate the modeling process. The approximate thin-filament equations provide a good deal of useful information about both the steady-state and dynamic behavior, and they have been utilized in applications. The question of model validation requires more discussion, however, and it is taken up again in the next chapter.

BIBLIOGRAPHICAL NOTES

The standard treatise on fiber spinning is

A. Ziabicki, *Fundamentals of Fibre Formation* (New York, NY: John Wiley, 1976).

The basic text on elongational flows is

C. J. S. Petrie, *Elongational Flows* (London: Pitman, 1979).

See also

C. J. S. Petrie and J. M. Dealy, in G. Astarita, G. Marrucci, and L. Nicolais, eds., *Rheology*, Vol. 1 (New York, NY: Plenum Press, 1980), p. 171.

The contents of this chapter are summarized in

M. M. Denn, "Continuous Drawing of Liquids to Form Fibers," *Ann. Rev. Fluid Mech.* **12** (1980) 365.

M. M. Denn, *"Fiber Spinning,"* in J. R. A. Pearson and S. M. Richardson, eds., *Computational Analysis of Polymer Processing* (London: Applied Science Publisher, 1983) p. 179.

See also

J. L. White, *Polymer Eng. Reviews* **1** (1981) 287.

The isothermal Newtonian thin filament equations are derived from the Navier-Stokes equations in

M. A. Matovich and J. R. A. Pearson, *Ind. Eng. Chem. Fundamentals* **8** (1969) 512.

W. W. Schultz and S. H. Davis, *J. Rheology* **26** (1982) 331.

The asymptotic analysis for the Maxwell fluid is in

M. M. Denn, C. J. S. Petrie, and P. Avenas, *AIChE J.* **21** (1975) 795.

For the finite element analysis of spinning, see

R. J. Fisher, M. M. Denn and R. I. Tanner, *Ind. Eng. Chem. Fundamentals* **19** (1980) 195.

R. Keunings, M. J. Crochet, and M. M. Denn, *Ind. Eng. Chem. Fundamentals* **22** (1983) 347.

The simulation of the pilot plant data is from

D. K. Gagon and M. M. Denn, *Polymer Eng. & Sci.* **21** (1981) 844,

where references to the sources of rheological and transport data are contained. The PET spinning data are reported in

H. H. George, *Polymer Eng. & Sci.* **22** (1982) 292.

Draw resonance is discussed in detail in the following review, which also treats other flow instabilities in polymer flows:

C. J. S. Petrie and M. M. Denn, *AIChE J.* **22** (1976) 209.

The first solutions, for an isothermal Newtonian fluid, were obtained independently by Kase and coworkers and by Pearson and Matovich; Kase's result was published in Japanese in 1966 and was not generally recognized in the west until much later. The earlier theory is contained in the results for viscoelastic liquids in

R. J. Fisher and M. M. Denn, *AIChE J.* **22** (1976) 236.
R. J. Fisher and M. M. Denn, *AIChE J.* **23** (1977) 23.
J. C. Chang and M. M. Denn, in G. Astarita, G. Marrucci, and L. Nicholais, eds., *Rheology*, Vol. 3 (New York, NY: Plenum Press, 1980) p. 9.

The first of these papers builds on an unpublished analysis by Zeichner,

G. R. Zeichner, "Spinnability of Viscoelastic Liquids," M. Ch.E. Thesis, University of Delaware, Newark DE, 1972.

There is a recent summary of outstanding problems in

M. M. Denn and J. R. A. Pearson, *Proc. 2nd World Congress Chem. Eng.* (1981), p. vi-354.

See also

S. Nam and D. C. Bogue, *Ind. Eng. Chem. Fundamentals* **23** (1984) 1,

who present additional data and discuss the limitations of the theory. Tsou and Bogue appear to have resolved problems associated with the proper initial conditions following spinneret flow in a paper to be published in *J. Non-Newtonian Fluid Mech.*

There has been a great deal of confusion in the spinning literature over the difference between sensitivity and stability, with the result that some utter nonsense has been published; see Denn and Pearson, cited above, for a further discussion. The literature on dynamic simulation and process dynamics is surveyed in the chapter in *Computational Analysis of Polymer Processing;* two treatments from a traditional process dynamics point of view are

S. Kase and M. M. Denn, *Proc. 1978 Joint Automatic Control Conference* (Pittsburgh, PA: Instrument Soc. of America, 1975), p. II-71.

M. M. Denn, "Modeling for Process Control," in C. T. Leondes, ed., *Control and Dynamic Systems,* Vol. 15 (New York, NY: Academic Press, 1979), p. 146.

The most recent work is

J. C. Chang, M. M. Denn, and S. Kase, *Ind. Eng. Chem. Fundamentals* **21** (1982) 13.

S. Kase and M. Araki, *J. Appl. Polymer Sci.* **27** (1982) 4439.

Kase and Araki show transfer functions for a large variety of input disturbances. It is perhaps significant that no application to real process disturbances has ever been reported, though some experimental work is underway at the time of writing.

13

MODEL VALIDATION

13.1 INTRODUCTION

Model validation is an essential component of any study, since the model must be shown to be reliable before it can be applied with confidence to new situations. (The goal of a modeling study *must* be application to new situations, in order to test the limits of understanding. There is little point in expending the substantial effort usually required, simply to obtain a means of interpolating within a region where behavior is already defined.) Model validation is usually seen as simply a comparison between some predictions and measurements, perhaps using tests of statistical significance. We shall not deal with the statistical aspect of model validation, which includes the problem of parameter estimation. Our concern here is with the basic concept of how one chooses the predictions that are to be made and the data that are to be taken.

13.2 STRUCTURAL SENSITIVITY AND TOVES

The model validation problem is nicely illustrated by Mary Hesse's *Tove syllogism*:

> Major premise: All Toves are white.
> Minor premise: My car is a Tove.
> Conclusion: Therefore my car is white.

This is total nonsense, of course, and the nonsense is transparent in this case: validation by observation that the car is white proves nothing about the "model" contained in the major and minor premises. Yet, had we replaced the two lines of gibberish by pages of equations and a multi-page computer code, the output of which predicted a white car—would we then feel so secure in rejecting the model?

A model must be probed; it must be asked questions whose answers can be verified. It is the responsibility of the modeler to seek out the questions that truly test the model. (An oracle who always predicts the rise and passage of the sun is reliable until he is asked about an eclipse.) We are not alluding here to parametric sensitivity of the type shown in Fig. 12.12, but rather *structural* sensitivity of the type seen in Figs. 12.9 and 12.14. Formulating the correct queries is not an easy task, and (save one that follows) we have no general principles to propose. Our purpose in this chapter is primarily to emphasize the significance of the problem, and to propose that even the most straightforward consideration of model validation requires concepts in the philosophy of science that are not within the experience of most engineers.

One important observation can be made, and it is again helpful to return to the Toves. The major premise here represents an axiom; the problems arise with the minor premise, which is a nonsensical statement about a very tangible object. In our modeling, the conservation principles play the role of the major premise; they are fundamental concepts that are taken as absolute. The constitutive equations collectively form the minor premise;* they are specific, often the result of narrow experimental studies, and open to question. The probing of a model is usually a probing of our constitutive hypotheses, and it is the constitutive hypotheses that provide the mechanism for exploring structural sensitivity.

13.3 COAL GASIFIER: SIMPLICITY

The agreement between the computed effluent and the plant observations in Table 7.1 is quite good, particularly in view of the fact that all parameters are taken from other data sources. Such agreement might be taken as validation of the model, were it not for the insight into the dominant physicochemical mechanisms contained in the kinetics-free analysis in Section 4.3, leading to the calculations in Table 7.2. The effluent calculations test noth-

---

*This is a bit too simple, since the simplifying approximations that might be applied to the constitutive equations are in the second category. The general notion seems useful, however.

ing in the detailed model except that the residence time is sufficiently long to allow steam and CO_2 to react with the carbon that is not to be oxidized. *Any reactor model with this characteristic must do a good job of representing the effluent, irrespective of what is predicted to occur within the reactor.*

In this particular case we have a bit of additional evidence to support the general structure of the model. The calculated optimum for the Illinois coal is in fact close to the actual conditions used for this coal in the Westfield tests, and the computed maximum temperature is just below the ash softening temperature. Thus, the model does seem to incorporate the actual operating experience on this reactor system, and it is on that basis that we feel some confidence in using it (with restraint!) to study other operating conditions and alternative designs. Nevertheless, adequate validation will be obtained only when we have temperature and composition measurements along the axis in the reaction zone, for here is where the many available models differ in their predictions; such measurements are not currently available, and it is not obvious that they can be obtained.

There are several important lessons here. The first, of course, reinforcing the discussion in Section 2.1, is that there is rarely a single model appropriate to all circumstances. The kinetics-free model requires considerably less computational effort than the detailed model, and is thus to be preferred in circumstances where we know the conversion of carbon and wish to compute the reactor effluent composition and temperature. A detailed model is required, however, if we wish to study questions of operability which involve unknown conversions and necessitate good estimates of the location and magnitude of the maximum temperature. Indeed, under some circumstances (e.g., hot standby operation, with a very small gas throughput) a two-dimensional detailed model is essential.

The other important point, which we shall expand upon subsequently, is that we have implicitly used the *principle of parsimony* (sometimes called *Occam's razor*) in the preceding paragraph. When two models fit the data equally well, we are to choose the simpler. Here, as is common in the engineering and science literature, we have taken *simpler* to mean *smaller number of computational steps*. This definition usually suffices, but we shall shortly see that it is inadequate.

13.4 FIBER SPINLINE: SIMPLICITY PARADOX

The comparison between the viscoelastic and Newtonian spinline models in Fig. 12.7 is a dramatic illustration of what we shall call the *simplicity paradox*. Given the inherent difficulty in making both the rheological and heat transfer measurements, it is difficult on the basis of absolute measurement accuracy alone to argue in favor of one or the other picture of the spinline: viscoelastic, with very small η/G, and a reliable heat transfer correlation from the literature; or negligible η/G, with the correct heat transfer coefficient being about 25% below that predicted by the Kase-Matsuo correlation.*

A viscoelastic spinline model for PET requires many more computational steps for any given spinning conditions than a Newtonian model does. There are at least two additional differential equations, and perhaps four or more, depending on the value of N; because η/G is small, the equations can become stiff in the numerical analysis sense and require specialized numerical integration techniques. If the number of computational steps is truly to be our measure of simplicity, then the principle of parsimony dictates that we select the Newtonian model and use it in our explorations of PET spinline operation.

This is a most unpleasant conclusion. We expect differences between Newtonian and viscoelastic fluid behavior as we modify operating policies; if we shroud the spinline from the quench air for a region just above the solidification point, for example (as in a patented process for partially oriented yarn), quite different stress patterns should result. The intuition of the experienced rheologist says that, measurement problems notwithstanding, PET *is* viscoelastic and extrapolation using the model should be on that basis. Shall we, then, abandon the principle of parsimony, which has long been one of the cornerstones of model evaluation, and follow intuition? If so, have we conceded that there is no real logic in the selection from among competing models, and that the ultimate determination is to be based on personal taste?

As with most paradoxes, there is a false premise. Here, the false premise is that the number of computational steps is an appropriate measure of relative simplicity. As we shall see in the next section, there is no essential conflict between our intuition and the principle of parsimony when we have abandoned our naive notions of what is "simple."

---

*We have an additional observation that suggests strongly that the value of G used in the simulations is approximately correct, for this is the value used in the stability calculations in Fig. 12.16. The stabilization at short spinline lengths is impossible unless G is roughly in the range used. The proper interpretation of these experiments has also been questioned, however (in hallway conversations at technical meetings, but not to our knowledge in the published literature, unfortunately), since substantial heat transfer and deformation might be taking place in the water bath, and thus the stabilization might be something quite different from that predicted by the theory. We shall thus disregard this additional evidence in the discussion at hand, though it is clearly relevant with regard to the subsequent intuitive argument for the viscoelastic model. In any event, this is nevertheless a good example of the additional discrimination afforded by an analysis of transients, particularly those predicting qualitative changes in behavior.

13.5 DEGREE OF FALSIFIABILITY

The quantification of the concept of simplicity has been discussed at length in the philosophy-of-science literature. This discussion has generally been in the context of the evaluation of scientific theories, and we shall play a bit fast and loose in carrying this discussion to the problem at hand and equating "model" with "theory." The obvious definitions of simplicity are easily dismissed (e.g. smallest number of axioms: axioms can always be combined). The most useful definition seems to us to be Popper's *degree of falsifiability*.

The starting point is the recognition that one never proves that a theory is correct (or validates a model!), for an infinite number of experiments would be required. Rather, one seeks the experiments that demonstrate that a theory is *incorrect*. The minimum number of experiments that need be performed to demonstrate that a theory is incorrect is a measure of the degree of falsifiability. The simpler a theory in this content, the lower the degree of falsifiability. We therefore take the degree of falsifiability to be the quantitative measure of simplicity. It is important to note that the ease of obtaining predictions from the theory (i.e. computational effort) never enters as a factor. (A model that does not contain the answer to a question is clearly simpler by this criterion than one that does, since no question need be asked to invalidate it. This points up the fact that only *relative* simplicity can be determined, since questions can be posed to any model for which responses cannot be provided. If model I is fully contained within model II, then I is simpler than II. The kinetics-free model is thus simpler than the detailed gasifier model, since the latter contains all of the information in the former.)

Let us now consider an experiment that we might carry out on melt spinning. We would choose a Newtonian fluid having a viscosity-temperature relation similar to that of PET, spin a filament, and measure the velocity profile. Three outcomes are possible:

1. The profile is that predicted by the Newtonian model, using the Kase-Matsuo correlation for h (Eq. 12.14).
2. The profile is that predicted by the Newtonian model, using a value of h close to 75% of that obtained from the Kase-Matsuo correlation.
3. Neither (1) nor (2) is true.

In case 1, we reject the Newtonian spinning model for PET, since that model requires that the heat transfer correlation be substantially in error. Similarly, in case 2 we reject the viscoelastic model, which requires that the heat transfer correlation be substantially correct. In case 3 we must reject both models. On the basis of this thought experiment it is evident that a single experiment is all that is required to invalidate either model of PET spinning, and hence both have a degree of falsifiability of unity. In this sense, then, they are equally simple.

Unfortunately, the experiment described here has never been carried out, and it is not obvious that we know how to do so at the present time.* Thus, while we have established that the models are equally simple, we do not have the means of discriminating on the basis of the critical experiment, and some other criterion must be employed. There is now no difficulty in introducing intuition, and even esthetics, since we are not violating the principle of parsimony. Thus, we may proceed with the use of the viscoelastic model.

13.6 CONCLUDING REMARKS

While Popper's concept of degree of falsifiability serves us well in resolving the simplicity paradox, it is not without its critics. Given the essential approximation in all theories, as well as experimental measurement error, no theory can be expected to agree perfectly with any single experiment, and thus all theories are rigorously falsified by strict application of the concept. Hence, judgment must enter with regard to the acceptability of agreement, and falsification becomes no more exact than validation. This point is well taken, but we nevertheless believe that overcoming the hurdle of the simplicity paradox is of overriding importance in the application of mathematical models, and we know of no better resolution for dealing with those cases in which a more intuitive measure like computational effort fails.

BIBLIOGRAPHICAL NOTES

For more on Toves, see Lewis Carroll's poem "Jabberwocky." The syllogism is from

> M. B. Hesse, *Models and Analogies in Science* (South Bend, IN: University of Notre Dame Press, 1966).

For a detailed discussion of the concept of simplicity, see Chapter 10 of

---

*Spinning a fiber from an inorganic glass appears to be a possibility, since inorganic glasses are Newtonian fluids. The experiment is confounded by the importance of radiative heat transfer in glass spinning, however, which is not a factor at the lower melt temperatures for polymers.

M. B. Hesse, *The Structure of Scientific Inference* (Berkeley, CA: University of California Press, 1974).

Karl Popper's ideas, which are most inadequately represented by this one paragraph summary, may be found in several of his books, including

K. R. Popper, *The Logic of Scientific Discovery* (New York, NY: Harper & Row, 1968).

One of the foremost critics of the Popper school is Thomas Kuhn, whose books are highly recommended for their insights into scientific thought:

T. S. Kuhn, *The Structure of Scientific Revolutions,* 2nd ed. (Chicago, IL: University of Chicago Press, 1970), especially Chapter 12.

T. S. Kuhn, *The Essential Tension* (Chicago, IL: University of Chicago Press, 1977), especially Chapter 13.

See also

S. Toulmin, *Human Understanding* (Princeton, NJ: Princeton University Press, 1972).

Toulmin's critique of Kuhn provides balance.

14

CASE STUDY: ACTIVATED SLUDGE PROCESS

14.1 INTRODUCTION

The activated sludge process for wastewater treatment was briefly described in Section 2.6 and shown schematically in Fig. 2.9; the heart of the process, the biochemical reactor and secondary clarifier, is shown schematically here in Fig. 14.1. We take it now as an example of the way in which the modeling tools that we have been developing can be utilized in the rational design and control of a complex process. It is important to emphasize that we are not aware of any wastewater plant that has ever been designed in this manner, but that is not an important issue. The steady-state model developed here leads to a plant design that is quite consistent with industry standards, and this is accomplished by using fundamental models for the most important pieces of equipment. Thus, the industry experience has been quantified in a model that can, with caution, be used to explore operating regimes beyond those normally encountered. Furthermore, because the steady-state model is expressed entirely in terms of equations, and uses none of the graphical techniques commonly encountered in the environmental engineering literature, hundreds of design alternatives, with complete economic evaluations for each, can be examined in minutes on a mainframe computer.

The discipline imposed by the modeling exercise points out some very important factors regarding design and operation. There are surprisingly few degrees of freedom available to the designer because of process and

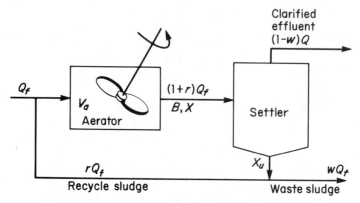

FIG. 14.1
Biochemical reactor and secondary clarifier in the activated sludge process.

effluent ("product") constraints; this is a general characteristic of process designs that is often overlooked. Design decisions having important economic consequences must be made on the weakest of foundations, yet research emphasis on the critical design issues is lacking here and has been

TABLE 14.1.
Typical composition of medium-strength domestic wastewater.

Components	Concentration, g/m^3*
Total BOD$_5$	200
Dissolved BOD$_5$	70
Total suspended solids (TSS)	200
Volatile suspended solids (VSS)	150
Total nitrogen (as N)	40
Ammonia nitrogen (as N)	25
Nitrite nitrogen (as N)	0
Nitrate nitrogen (as N)	0
Organic nitrogen (as N)	15
Total phosphorus	10
Dissolved phosphorus	7

*The unit g/m^3 is equivalent to mg/liter and parts per million (ppm), all of which are in common use.

14.2 PROCESS CONDITIONS 279

almost entirely elsewhere, where the economic consequences are far less significant. The process dynamics depend critically on the operation of the secondary clarifier, and time constants can change by *orders of magnitude* because of small changes in the operation of the clarifier. The aeration system, which is a major factor in the economics, will often be underdesigned if dynamics are not taken into account. Finally, as will often be the case in process design, the economic cost surface is quite insensitive over a large range of design options, and controllability becomes an important factor in the selection of a design.

The structure of this chapter differs from that of the other case studies, since we are concerned with the entire process. The models for the industrial units are developed in a series of appendixes, with the emphasis on the three units having the greatest importance in the economics and the operation: the biochemical reactor (aeration basin), the secondary clarifier, and the primary settler. (These models cover the range from fundamental to empirical. The reactor model differs little from those studied before, and thus requires little attention. The theory underlying the secondary clarifier is developed in some detail.) The body of the chapter deals with the *use* of these unit models in the analysis of an integrated process.

14.2 PROCESS CONDITIONS

We will be simulating a plant that is to treat 20,000 m^3/day of medium-strength wastewater. This is an intermediate size plant, which can be expected to experience significant diurnal variation in flow. The average composition of the medium-strength wastewater is shown in Table 14.1. The diurnal variations shown in Fig. 14.2 are typical, and are used in the dynamic simulations here.

The effluent standards used for the simulations are those proposed in 1977 by the Water Pollution Control Federation: total effluent BOD_5 (dissolved and suspended) of 20 g/m^3, and total effluent suspended solids (TSS) of 20 g/m^3. (These are two-thirds the levels required by U.S. Public Law 92-500, but the stricter standards seemed more in accord with current thinking when this simulation was done in 1979 and 1980.) The effluent BOD will be made up in part of suspended material, mostly sludge, that has been carried over in the clarifier effluent; the 20 g/m^3 TSS consists of approximately 16 g/m^3 volatile suspended solids (VSS), which are organic.*

---

*The composition of the cells which is used for all stoichiometric computations involving VSS is $C_5H_7NO_2P_{0.1}$. This is close to that of typical microorganisms. Five moles of oxygen are required to convert one formula weight to products including CO_2 and water.

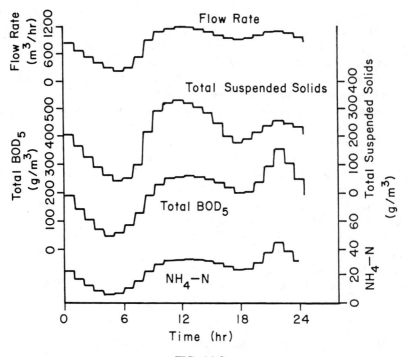

FIG. 14.2.
24-h time series of influent flow rate and wastewater characteristics (adapted from Metcalf and Eddy, 1972).

Analysis of the model equations developed in the appendices brings out an important feature of the steady-state model for the plant: For a given average wastewater flow and wastewater characteristics, the combination of effluent composition constraints and plant topological structure leaves the designer only *three degrees of freedom!* These are somewhat arbitrary as long as independent choices are made, but we have found it convenient to use *solid removal in the primary settler, sludge age,* and *solids level* (VSS) *in the aeration tank.* When these three design parameters are set, the entire plant is specified. This small number of degrees of freedom obviously makes interactive computation for the consideration of design alternatives most attractive.

14.3 PROCESS ECONOMICS

The total cost of the process includes the cost of construction, plus the anticipated cost of operation and maintenance (O&M) over the expected life-

14.3 PROCESS ECONOMICS

time of the plant. Unit construction and operating costs typically depend on design variables like volume or throughput. Costs have been published specifically for the units in wastewater treatment processes, and process industry cost functions have also been published. These data generally follow a power function relation (in contrast to the linear functions assumed in Section 4.3); the installed cost of the primary clarifier is estimated from the following equation, for example:

Construction cost (1978 U.S. dollars)
$$= 3355 \times \text{(Area of primary clarifier)}^{0.67} \quad \textbf{(14.1)}$$

Data of this type are typically accurate only to within 15 to 30%, so all unit process costs in the simulations described here were rounded to two significant figures. Several indices exist for adjusting cost figures like those in Eq. (14.1) to provide an up-to-date economic data base.

An accounting method must be chosen in order to compute costs. The method selected here is that of *net present worth*, which computes the capital that must be in hand at the start of the project in order to pay for construction and the annual operation and maintenance costs for the duration of the project. Though it is outside the scope of our general subject, it is useful to develop the equations for the net present worth method (as an elementary exercise in economic modeling, perhaps), since this type of analysis is not commonly covered in many curricula.

We need to estimate the average interest and inflation rates over the plant lifetime of N years; let i and I represent these rates, respectively. P_n represents the interest-earning capital available in year $n - 1$ (that is, P_0 is available for the first year), and C_n the O&M cost in that year. We then have the following two equations to solve:

$$P_{n+1} = (1 + i)P_n - C_n \quad \textbf{(14.2a)}$$

$$C_{n+1} = (1 + I)C_n \quad \textbf{(14.2b)}$$

This set of linear difference equations must be solved with two boundary values: C_0, the O&M cost for the first year is given; and $P_N = 0$, since no further capital is required. The solution to Eq. (14.2b) is written immediately as

$$C_n = C_0(1 + I)^n \quad \textbf{(14.3)}$$

The solution to Eq. (14.2c) consists of the sum of a homogeneous and a particular part; the former is

$$\text{homogeneous:} \quad P_n = \text{constant} \times (1 + i)^n \quad \textbf{(14.4a)}$$

The particular solution is found using the method of undetermined coefficients to be

$$\text{particular:} \quad P_n = \frac{C_0}{i - I}(1 + I)^n \tag{14.4b}$$

The constant is evaluated from the condition $P_N = 0$, giving a solution

$$P_n = \left[\frac{(1 + I)^n(1 + i)^N - (1 + I)^N(1 + i)^n}{(i - I)(1 + i)^N}\right] C_0 \tag{14.5}$$

We are interested in the value P_0, since this represents the starting capital. This value is

$$P_0 = \frac{1}{i - I}\left[1 - \left(\frac{1 + I}{1 + i}\right)^N\right] C_0 \tag{14.6}$$

The quantity multiplying C_0 is known as the *present worth factor*, PWF; since we normally expect $I \ll 1$, it is closely approximated by

$$\text{PWF} \approx \frac{1}{i - I}\left[1 - \frac{1}{(1 + i - I)^N}\right] \tag{14.7}$$

Thus, PWF depends on the *net* interest rate (the difference between true interest and inflation). For large N, PWF is slightly less than the reciprocal of the net interest rate.

The total cost is equal to the capital required for construction, plus PWF multiplied by the O&M costs for the first year. All of the calculations shown here are for a net interest rate of 10% and a lifetime of 20 years. This corresponds to a PWF of 8.5.

14.4 BASE CASE

The model equations were first used to compute a base-case design conforming to usual practice, with 60% solids removal in the primary settler, a 10-day sludge age, and aeration basin VSS of 2000 g/m^3.

The computed costs are shown in Tables 14.2 through 14.4 The total present cost in 1978 U.S. dollars is $5.3 million, $2.9 million construction and $2.4 million O&M. This large O&M factor points up the danger of basing cost estimates on construction only. If, in fact, the net interest rate

14.4 BASE CASE

were dropped to 5%, the PWF would increase to 12.5 and the O&M costs to $3.5 million. The costs are clearly very sensitive to interest rates, and lower rates will favor capital cost-intensive plants with low O&M costs.

The air supply system warrants particular attention. The design basis is to use *firm blower capacity;* this is blower capacity required to meet the design standard with one blower out of service. The air supply is a major factor in both the construction and O&M costs. It represents most of the plant electrical costs, for example, which are in turn 13% of the total O&M cost. We will find subsequently, with the dynamic simulations, that an air system sized in this way is just adequate to meet the peak oxygen demand during periods of high flow and BOD loading, and only if all blowers are

TABLE 14.2.
Construction costs for base case.

	Cost (10^3 U.S. Dollars)	Percentage of Total
Preliminary treatment	130	5.5
Primary clarifier	350	14.8
Primary sludge pump	20	0.8
Aeration tank	310	13.1
Air supply system	290	12.3
Secondary settler	310	13.1
Recycle pump	46	1.9
Wastage pump	2	0.1
Chlorination tank	50	2.1
Chlorine feed system	6	0.3
Sludge thickener	39	1.7
Thickener pump	2	0.1
Anaerobic digester	180	7.6
Solids centrifuge	140	5.9
Laboratory facilities	150	6.4
Garage and shop	46	1.9
Yardwork	290	12.3
Total construction cost	2362	100.0
Land (18.6 acres)	93	
Engineering	270	
Legal, fiscal, administration	35	
Interest during construction	150	
Total construction capital	2910	

TABLE 14.3.
First-year operation and maintenance costs.

	Cost (10^3 U.S. Dollars)					
	Operation	Maintenance	Material and Supply	Electrical	Chemical	Total	Percentage of Total
Preliminary treatment	10	4.9	5.6	0	0	20.5	7.3
Primary clarifier	7.6	3.9	3.2	0	0	14.7	5.3
Primary sludge pump	2.5	0.8	0.7	0.04	0	4	1.4
Aeration tank	0	0	0	0	0	0	0
Air supply system	29	6.3	8.2	36	0	79.5	28.4
Secondary settler	7	3.6	2.8	0	0	13.4	4.8
Recycle pump	2.2	1.7	0.5	1.3	0	5.7	2
Wastage pump	0.4	0.4	0.01	0.02	0	0.8	0.3
Chlorination tank	0	0	0	0	0	0	0
Chlorine feed system	4.8	0.9	0	0	10	18.1	6.5
Sludge thickener	1.3	0.7	0.3	0	0	2.3	0.8
Thickener pump	0.4	0.4	0.01	0.03	0	0.8	0.3
Anaerobic digester	3.1	2.1	2.5	0	0	7.7	2.8
Solids centrifuge	16	2.3	9.8	0	9	37.5	13.4
Laboratory facilities	23	1.2	1.3	0	0	25.5	9.1
Yardwork	0	9.6	2.2	0	0	11.8	4.2
Administration and general	8.1	1.7	6.1	0	0	15.9	5.7
Indirect labor costs	16	5.6	0	0	0	21.6	7.7
Total	131.5	45.9	45.6	37.4	19.4	279.7	100.0
Percentage of annual costs	47.0	16.4	16.3	13.4	6.9	100.0	

TABLE 14.4.
Economic summary for base case, $N = 20$ years, $i - I = 0.10$.

Construction capital (10^6 U.S. dollars)	2.9
First-year O&M (10^6 U.S. dollars)	0.28
PWF	8.5
Present worth of 20 years O&M (10^6 U.S. dollars)	2.4
Present total cost (10^6 U.S. dollars)	5.3

in service. Thus, the simulations suggest that a more conservative design basis be employed for the air supply system, perhaps one of specifying a firm blower capacity adequate to meet the expected maximum oxygen demand. Such a modification would have an impact on the total system costs and could affect the selection of a design.

14.5 DESIGN SENSITIVITY

There are only three degrees of freedom available in the plant model, so an examination of design sensitivity is straightforward. The fractional solids removal in the primary chapter was examined first over a range of sludge ages and aeration basin VSS for wastewaters of various strengths, with and without polymer addition to enhance settling through flocculation. A typical computation is shown in Fig. 14.3 for $\theta = 10$ days and reactor VSS = 2000 g/m^3; the base case is the 60% removal point on this curve. The cost is quite insensitive to primary removal in the range from 10% to 50%, but the minimum cost occurs with no primary removal at all. This is because the savings in aeration basin and air supply costs resulting from the smaller BOD load to be treated are more than offset by the cost of installing the primary clarifier and the additional load on the anaerobic digester. The same result was obtained in all cases, though the penalty decreased with increasing wastewater strength.

The suggestion has been made in the literature, based on operating plant cost data, that the primary clarifier is not required. The simulations using the plant model confirm this, but two caveats are in order. First, it must be recalled that the kinetic data were inadequate to account for the possibly slower reaction rate of suspended BOD relative to dissolved BOD. Second, the design equation for sizing the primary clarifier is empirical and derived from data having considerable scatter. One of the most important

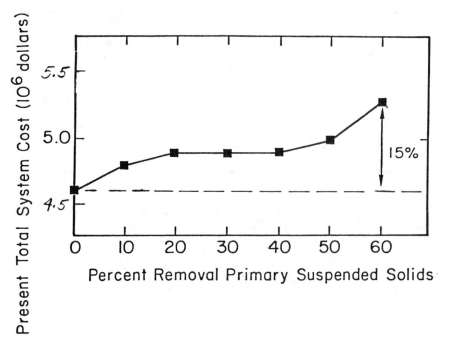

FIG. 14.3.
Cost as a function of primary solids removal efficiency, $\theta = 10$ days and reactor VSS = 2000 g/m^3 (Paterson and Denn, 1980, 1983, copyright Elsevier Sequoia, reproduced with permission).

results of the modeling study is to point to the significant economic motivation to improve the level of fundamental understanding of these two areas. Finally, despite the increased cost, there may be some motivation to include a primary clarifier. Even with a relatively short detention time, say one hour, the primary clarifier will tend to spread the effect of feed variations, particularly shock loads. In the event of an aeration system failure, the presence of the primary unit will ensure that at least minimal treatment is provided to the wastewater before it is discharged to the receiving water body. On that basis, a primary clarifier that removes about 40% of the influent solids would be suggested to the extent that the model is reliable.

The detailed exploration of the effect of sludge age and aeration basin-suspended solids was carried out only for the case of no primary solids removal, since that always represented the minimum cost. The computed cost and the recycle ratio to the aeration basin are shown in Fig. 14.4 for a sludge age of 10 days. The recycle ratios are very large for high values of the reactor-suspended solids, exceeding unity in some cases. A large re-

14.5 DESIGN SENSITIVITY

cycle rate in a plant with a high microorganism concentration (and hence a small aeration basin) results in a very short hydraulic residence time in the aeration basin. This could lead to operability problems during transients. Indeed, conventional activated sludge plants do not normally operate with recycle ratios greater than 0.50. The calculations shown in Fig. 14.4 include the effect of constraining r to be less than 0.50.

The decrease in cost as the VSS increase from 1000 to 2500 g/m^3 is a consequence of the reduced volume of the aeration basin, since the basin volume is inversely proportional to reactor suspended solids. This trend is ultimately offset by the increased capital and operational costs of the air supply system and the recycle pump. The very rapid cost increase for the

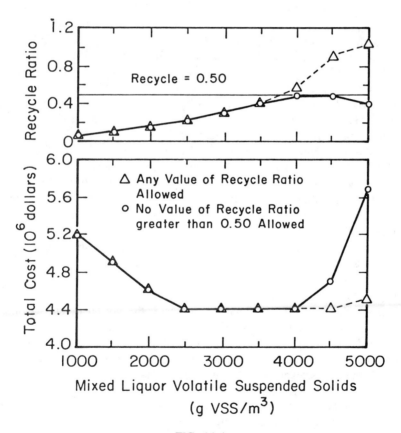

FIG. 14.4.
Cost and recycle ratio as a function of reactor suspended solids, $\theta = 10$ days and no primary solids removal (Paterson and Denn, 1980, 1983, copyright Elsevier Sequoia, reproduced with permission).

constrained solution when the suspended solids level is greater than 4000 g/m^3 is associated with the increased cost of the secondary clarifier. Because of the constrained recycle ratio, a clarifier sized on the basis of effluent suspended solids cannot satisfy the thickening constraints, and the area must be increased in order to pass the required amount of solids back to the aeration basis. The increased area will lead to better clarification as well, so the effluent quality will be better than the design value of 20 g/m^3, but at the expense of a large cost increase.

Plant costs for VSS from 1000 to 5000 g/m^3 and sludge ages from 2 to 46 days are summarized in Fig. 14.5. It is striking to note the relative insensitivity of the cost over wide changes in design variables. There are two least-cost designs, both having the same total present system cost of $4.3 million. These plants both have VSS of 4000 g/m^3; one has a sludge age of 12 days and the other of 14 days. A very large region, however, including sludge ages from 6 to 36 days and VSS from 2500 to 4500 g/m^3, has a cost of only $4.4 million, which is well within the uncertainty of the calculation. Plants with a total cost that is no more than 10% above the minimum exist for sludge ages from 2 to 38 days and VSS from 1000 to 4500 g/m^3. The minimum cost plants have particularly high recycle ratios and might therefore be expected to be dynamically more sensitive to disturbances than are adjacent lower recycle designs of slightly higher cost.

All the calculations described above were carried out using the sludge-settling data in Fig. 14.A5 (see Appendix to this chapter). Comparable calculations were carried out using other sludge-settling functions. In one, the initial settling velocity was assumed (unrealistically) to continue to increase with increasing sludge age beyond 15 days. The continuously improving ability to settle does not result in any further decrease in cost, because the secondary clarifier is sized by clarification requirements for sludge ages greater than 12 days. This result emphasizes the importance of the clarification function in establishing the system design and the fact that there is a point of zero marginal return from improved settling.

As the preceding calculations indicate, the notion of an "optimal" design is somewhat nebulous. The minimum cost designs can be clearly identified, and these must be the standard against which other designs are evaluated. There is a large range of designs with costs just slightly above the minimum, however, as illustrated in Fig. 14.5, and the cost difference is probably within the uncertainty even of a calculation with good kinetic and settling data on the sludge system in question. Hence, the designer has considerable flexibility and can use other factors besides cost in making a final selection.

Plant controllability should be a primary factor in selecting from among alternatives, and the remainder of the chapter is concerned with this subject. Two designs were selected for detailed study, either of which might be con-

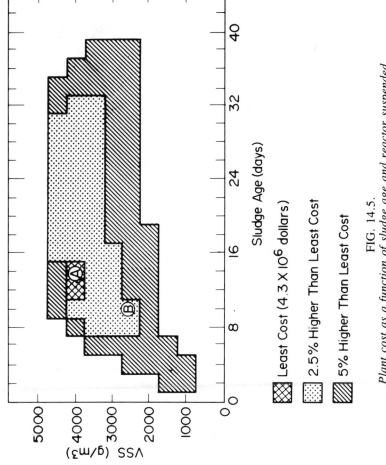

FIG. 14.5.

Plant cost as a function of sludge age and reactor suspended solids (Paterson and Denn, 1980, 1983, copyright Elsevier Sequoia, reproduced with permission). Plants A and B are studied subsequently for dynamical response.

sidered "optimal" on the basis of the steady state calculations. Plant A is one of the two least-cost designs. It has a sludge age of 14 days, VSS of 4000 g/m^3, and a recycle ratio of 0.45. This plant represents a departure from usual practice in both the high concentration of suspended solids and the large recycle ratio. Plant B has a total present system cost of $4.4 million, only 2% higher than the minimum cost, with a sludge age of 10 days, VSS of 2500 g/m^3, and a recycle ratio of 0.22. The lower solids level and the lower recycle ratio relative to plant A could be expected to result in a longer aeration basin residence time and thus improved dynamic stability. The costs of both plants are to be contrasted with $5.3 million for the base case design to treat the same wastewater and to meet the same effluent standards.

14.6 DYNAMICS AND CONTROL

14.6.1 Recycle Systems

The presence of a recycle loop, as in Fig. 14.1, can cause large changes in the dynamical response of a process. This phenomenon seems to have been little studied, but standard methods of frequency analysis can be used to show that the plant may exhibit much greater dynamical sensitivity and longer response times than the individual elements. We first came upon this phenomenon (though it had been observed earlier) when trying to interpret dynamical simulations of the aeration-basin–secondary-clarifier system following step changes in the feed flow rate, and we shall discuss it here only in that context.

We shall refer frequently to the *height of the sludge blanket*. This point is defined visually in clarifier operation as the level at which the concentrated sludge layer ends; we define it in the context of the clarifier model described in the Appendix as the interface between the concentrated and dilute zones in the thickening section. An *underloaded* settler is one in which the flux is sufficiently low that there is no concentrated zone in the thickening section; an *overloaded* settler is one in which the feed flux is too high, solids cannot be passed through rapidly enough, and the sludge blanket rises up into the clarification zone. Secondary clarifiers are reported sometimes to run in an underloaded mode as a protection against sudden large influxes. An overloaded unit is to be avoided, since the effluent solids level is likely to increase under those conditions.

In developing control strategies for an activated sludge process a primary goal must be the prevention of large excursions in the aeration basin VSS, since this could lead to changes in the microorganism population that

14.6 DYNAMICS AND CONTROL

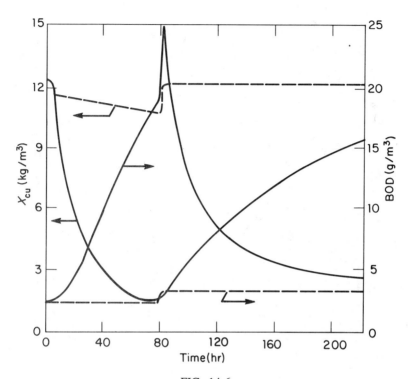

FIG. 14.6.
Simulations of a reactor-secondary clarifier system to step changes in influent flow rate at 2 and 80 h. Solid line: recycle ratio control. Dashed line: constant recycle flow rate (after Attir and Denn, 1978, copyright American Institute of Chemical Engineers, reproduced with permission).

would affect kinetic rates and settling characteristics. It is possible to show through an analysis of the equations in Section 14.A1, together with some simplifying assumptions about settler dynamics, that this goal should be achieved with a policy of ratio control on the recycle; that is, the recycle flow rate rQ_f is always kept in a constant ratio to the influent rate Q_f ($r =$ constant). There is also a strong incentive to keep the underflow rate $(r + w)Q_f$ from the secondary clarifier at a constant value, since this should help to avoid settler upsets. Clearly there is an essential conflict that prevents simultaneous implementation of these two policies unless stored sludge is available for recycle when needed ($w < 0$).

Some simulations implementing these two control policies are shown in Fig. 14.6. The kinetic constants are different from those used elsewhere in this chapter, but that is an unimportant detail. The aeration basin residence time is 3.3 hr, and the solids residence time in the secondary clarifier is less

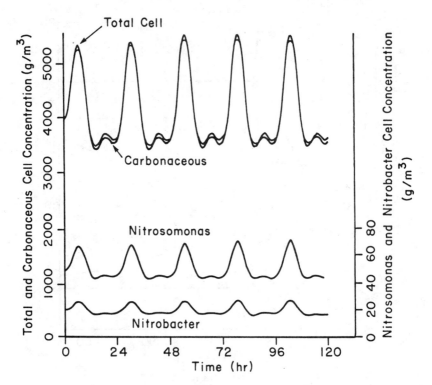

FIG. 14.7.
Aeration basin response, plant A, constant clarifier underflow and recycle (Paterson and Denn, 1980, 1983, copyright Elsevier Sequoia, reproduced with permission).

than 6 hr; 6 hr is the time required for convective flow, without accounting for the increased solids rate because of settling. The sludge blanket height was set at 0.6 m at $t = 0$; this low level was chosen in order to facilitate illustration of a point, but a low sludge level might be maintained to minimize solids residence time in order to avoid organism death in the settler, as well as to provide space for the sludge blanket to rise in the event of a rapid influx.

The initial flowrate is 200 m^3/hr. At time $t = 2$ hr there is a 25% step reduction in the flow rate to 150 m^3/hr; at $t = 80$ hr there is a step increase to the original 200 m^3/hr. The solid lines in Fig. 14.6 show the response when the system is maintained under recycle ratio control, with $r = 0.2$. The sludge level falls rapidly, and after a short time the settler is underloaded. The effect on the dynamics is dramatic; steady state has not been reached 120 hr after the step increase in flow rate.

14.6 DYNAMICS AND CONTROL

The dashed lines in Fig. 14.6 show the same case with a single change. Here, the recycle flow rate was maintained at the steady-state value of 40 m³/hr at all times. The effect is to recycle slightly more solids and to prevent serious underloading in the settler. The approach to steady state following the step increase is rapid, despite a difference of less than 6% in reactor residence time in the two cases.

The essential difference between these two cases is the existence of a limiting layer in the settler. Such a layer acts as a buffer and allows the settler underflow concentration to remain relatively constant in time. This prevents reactor disturbances from being fed back and amplified. It is evident from this simulation result that *a primary goal of a control strategy must be to prevent settler underloading* because of the possible adverse effect on dynamics and the ability of the system to recover from upsets. The presence of a limiting layer effectively uncouples the reactor dynamics from those of the settler, eliminating the unfavorable effect of recycle on system dynamics.

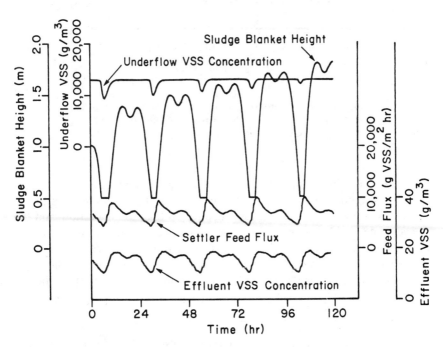

FIG. 14.8.
Secondary clarifier response, plant A, constant clarifier underflow and recycle (Paterson and Denn, 1980, 1983, copyright Elsevier Sequoia, reproduced with permission).

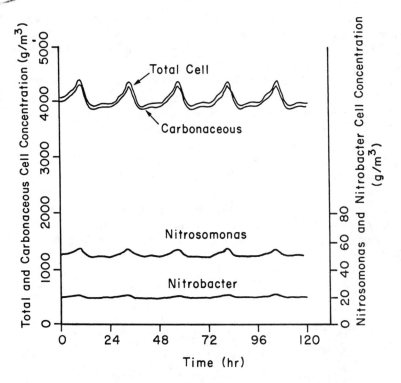

FIG. 14.9.
Aeration Basin response, plant A, ratio control on clarifier underflow and recycle (Paterson and Denn, 1980, 1983, copyright Elsevier Sequoia, reproduced with permission).

Further simulations* with a 24-hr cyclic forcing establish that a reasonable compromise between reactor and sludge blanket control can be effected by a policy that utilizes ratio control on both the underflow and the recycle rate (and hence on the wastage rate). Sludge storage is not required, the sludge blanket height varied only slightly in the test cases, and reactor solids fluctuations were reduced to 30% of those in the uncontrolled systems. This policy was therefore one of those used in the full plant simulations.

---

*These followed a classical steady-state control analysis, in which the unique underflow rate was computed that would keep the sludge blanket height fixed for each flow rate, assuming that the system was at steady state. In all cases the ratio of underflow to influent rate was within one percent of a constant value.

14.6.2 Plant Dynamics

Simulations were carried out for plants A and B, defined in Section 14.5, for the inflow variations shown in Fig. 14.2, using a variety of control strategies. It is to be recalled that plant A is the minimum-cost design, and hence the "optimum" in the classical sense; plant B is slightly more costly, but might be expected to exhibit better operability because of the lower recycle ratio and aeration basin solids loading.

The 5-day dynamic response of plant A with no control (constant clarifier underflow and recycle) is shown in Figs. 14.7 and 14.8. The plant experiences wide diurnal variations in both aeration basin cell concentration and sludge blanket height. Figures 14.9 and 14.10 illustrate the 5-day dynamic response when ratio control is used on the secondary clarifier underflow and sludge recycle flow rates. The diurnal variations in both aeration

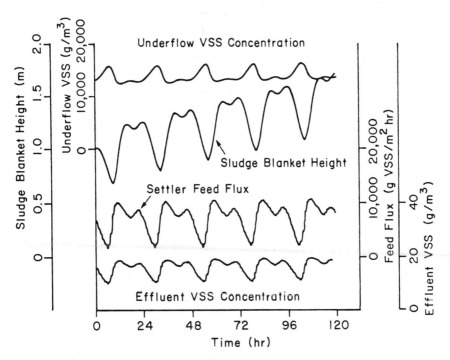

FIG. 14.10.
Secondary clarifier response, plant A, ratio control on clarifier underflow and recycle (Paterson and Denn, 1980, 1983, copyright Elsevier Sequoia, reprinted with permission).

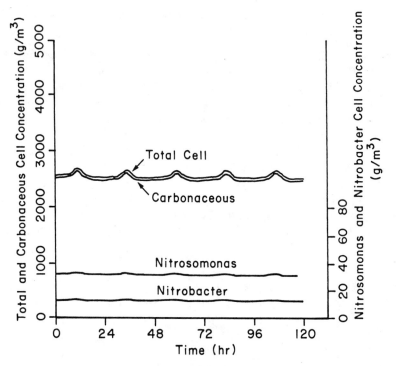

FIG. 14.11.
Aeration basin response, plant B, ratio control on clarifier underflow and recycle (Paterson and Denn, 1980, 1983, copyright Elsevier Sequoia, reprinted with permission).

basin cell concentration and sludge blanket height have been reduced considerably, though the sludge blanket continues to rise.

The best comparison of the relative operability of plants A and B would be based on the performance of each plant under ratio control. Figures 14.11 and 14.12 show the 5-day dynamic response of plant B using ratio control. The diurnal variations in the aeration basin cell concentrations are significantly less for plant B than for plant A. Similarly, the overall stability of the sludge blanket is superior for plant B. Thus, the dynamic simulations indicate that the minimum-cost plant A is dynamically more sensitive, and hence more difficult to control, than the more conventional plant B. Given the small difference in computed cost between the two plants, plant B appears to be a better choice of a final design.

The final effluent quality for plant B is shown for the case where no control is used (constant underflow and recycle) in Fig. 14.13, and in Fig.

14.14 when ratio control is used. The effluent quality for both cases is almost identical, and is also indistinguishable from the effluent of plant A both with and without control. The invariance of the effluent quality emphasizes an important fact: *Process control in response to diurnal feed variations serves to control the internal variables and to maintain the operability of the process, but it has little or no effect on effluent quality, which is fixed by the design and the mean flows.* Thus, transient effluent quality should not be used as a basis of comparison between control systems in a well-designed plant.

14.6.3 Oxygen Control

The kinetic expressions used in the steady-state and transient models require that the dissolved oxygen level always be above the minimum value at which

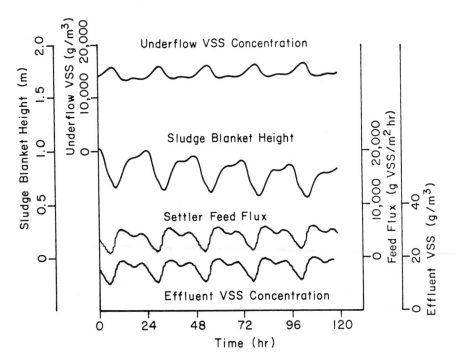

FIG. 14.12.
Secondary clarifier response, plant B, ratio control on clarifier underflow and recycle (Paterson and Denn, 1980, 1983, copyright Elsevier Sequoia, reprinted with permission).

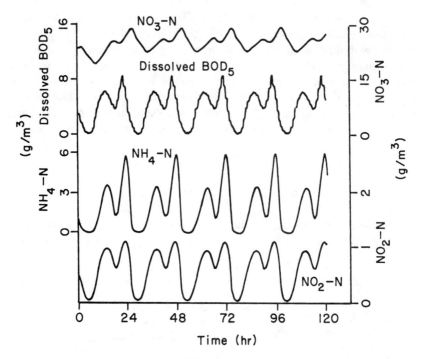

FIG. 14.13.
Final effluent quality, plant B, constant clarifier underflow and recycle (Paterson and Denn, 1980, 1983, copyright Elsevier Sequoia, reprinted with permission).

kinetics are oxygen independent. The air supply system is sized to ensure that this is true on the average, but the dissolved oxygen concentration will vary during the transient unless control is used. The computed dissolved oxygen concentration during a 24-hr transient is shown in Fig. 14.15; essentially the same curve is obtained for plants A and B, with and without ratio control on the underflow and recycle. This is because the rate of oxygen uptake is determined by the conversion rate of BOD, which is nearly the same in all cases.

Negative values of the computed dissolved oxygen concentration, as shown in Fig. 14.15, serve as an indication of the extent to which the oxygen demand is exceeding the oxygen supply. This is an artifact of a limitation in the model. The dissolved oxygen concentration obviously can never drop below zero; as the dissolved oxygen concentration approaches zero, the metabolic activity of the bacteria drops off rapidly. This dependence of the rates

14.6 DYNAMICS AND CONTROL

on the dissolved oxygen concentration was not included in the dynamic model equations, and was never added later because of the recognition that oxygen control was needed in any event.

Several alternative dissolved oxygen control strategies were evaluated using the dynamic model. The only successful strategy requires continuous dissolved oxygen measurement. (In practice, since the integration scheme uses finite time steps, an oxygen measurement is required about once every 15 min.) The result in Fig. 14.16 shows the dissolved oxygen concentration for the simulation of plant B when the blower rate is increased or decreased by 10% following each dissolved oxygen measurement in which the concentration is found to be below 2.0 g/m^3 or above 3.0 gm^3, respectively. This is a simple and effective scheme as long as reliable dissolved oxygen measurements are available.

The maximum air flow rate for the control shown in Fig. 14.16 is approximately twice the design air flow rate (the firm blower capacity). Some

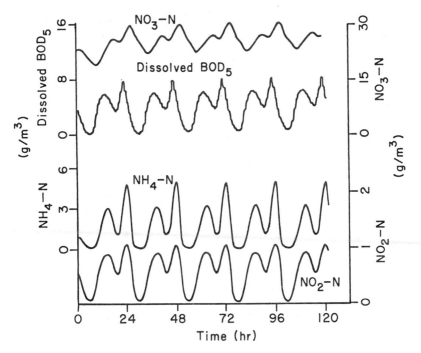

FIG. 14.14.
Final effluent quality, plant B, ratio control on clarifier underflow and recycle (Paterson and Denn, 1980, 1983, copyright Elsevier Sequoia, reprinted with permission).

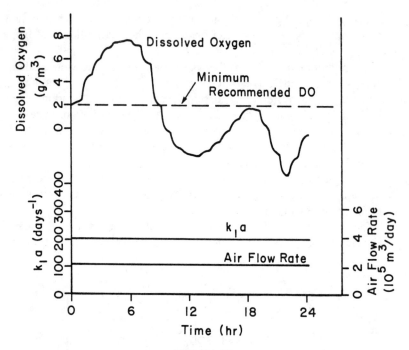

FIG. 14.15.
Aeration basin dissolved oxygen with constant air flow rate, plant B with ratio control on underflow and recycle (Paterson and Denn, 1980, 1983, copyright Elsevier Sequoia, reproduced with permission).

small and medium-sized plants have a maximum blower capacity with all units in service that exceeds the firm blower capacity by a factor of 1.5–2.0, so it might be possible to implement the oxygen control with the blowers designed by the steady-state program as long as all blowers are usually in working condition. A more conservative strategy would be to design the process with a firm blower capacity based on the maximum demand.

14.7 CONCLUDING REMARKS

This case study points up several interesting factors. First, it shows how relatively simple modeling can lead to improved understanding of the manner in which a process operates. The important interplay between design and

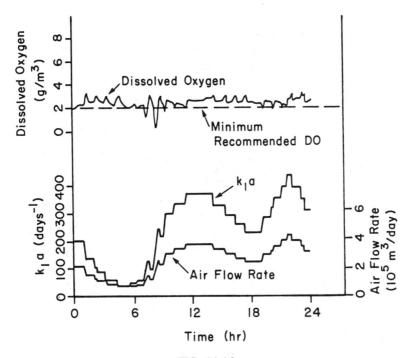

FIG. 14.16.
Aeration basin dissolved oxygen with oxygen feedback control, plant B with ratio control on underflow and recycle (Paterson and Denn, 1980, 1983, copyright Elsevier Sequoia, reproduced with permission).

control, for example, has only recently begun to receive serious discussion in the process engineering literature, and little if any in the environmental engineering literature. The issues raised regarding primary clarification, the effect of recycle on operability, and the need for good oxygen measurement and control are new in this application; some have been recognized by practioners, but without the logical foundation afforded by a reliable model which permits exploration of sensitivities and alternatives.

The other important factor is the illustration of how the discipline imposed by the modeling effort establishes the alternatives open to the designer and, more important, those areas needing further study. The clarification function in the primary and secondary units has a major impact on plant economics, and yet it is hardly understood at all. Here is where research expenditures need to go!

APPENDIX

14.A1   BIOCHEMICAL REACTOR (AERATION BASIN)

The aeration basin is a multiphase system, containing wastewater with dissolved material, microorganism flocs and suspended solids, and air bubbles from which oxygen is transferred to the liquid. The reactor is traditionally treated as though it contained a homogeneous liquid with physical properties characteristic of the suspension. The biochemical reactions are treated as homogeneous reactions, since the reaction sites (i.e., the microorganisms) are assumed to be well dispersed throughout the reactor such that the system appears to be homogeneous on a local scale and mass transfer into the floc is thought not to be limiting. All oxygen-utilizing chemical species are lumped into a single pseudo-species that is measured in standard oxygen uptake tests as *biological oxygen demand,* or BOD (sometimes BOD_5, since the standard test measures 5-day oxygen uptake). BOD kinetics for particular carbon-utilizing microorganisms can be readily measured; all such organisms are lumped into a single pseudo-species denoted *carbonaceous microorganisms.* The concentration of BOD is denoted B, and that of the carbonaceous microorganisms X_c; this nomenclature deviates from that used elsewhere in the text, but it should cause no confusion and facilitates comparison with the published literature.

   The reaction vessels are taken to be well stirred, with the nomenclature as shown in Fig. 14.1. The density of a suspension will be linear in suspended solids, and we assume that BOD levels have little effect on the liquid density, so the assumption of linearity is adequate and only volumetric flow rates need appear in mass balances (cf. Section 4.2). The equations

14.A1 BIOCHEMICAL REACTOR (AERATION BASIN)

for BOD and carbonaceous organisms are then

$$V_a \frac{dB}{dt} = Q_f(B_f - B) - r_B V_a \quad \text{(14.A1)}$$

$$V_a \frac{dX_c}{dt} = rQ_f X_{cu} - (1+r)Q_f X_c + r_c V_a \quad \text{(14.A2)}$$

Concentrations are measured in grams per cubic meter (g/m^3), which is equivalent to parts per million and milligrams per liter.

The microorganism growth rate, r_c, is usually represented by a three-parameter form that accounts for an inhibiting effect of excess substrate (i.e., food), and includes the fact that the growth rate of a population must always be proportional to the size of the population:

$$r_c = \left(\frac{\mu_{mc} B}{K_B + B} - b_c \right) X_c \quad \text{(14.A3)}$$

μ_{mc}, K_B, and b_c are parameters that will depend on the particular system. The first term in parentheses is associated with BOD uptake and conversion of the carbon-containing species to CO_2 and water, while the second represents *endogenous respiration;* the latter is metabolism of cell mass to provide energy for various cell functions. The organism growth depends on the availability of sufficient dissolved oxygen. Growth rates appear to be insensitive to oxygen concentration as long as it is in excess of about 2 g/m^3, but they drop off rapidly at oxygen levels below that value. It is assumed in all reactor calculations that the air supply system is adequate to maintain the dissolved oxygen at a level at which oxygen need not appear in the kinetic expression.

BOD conversion is related through a stoichiometric constant, denoted Y_c, to the first of the terms in Eq. (14.A3):

$$r_B = \frac{1}{Y_c} \frac{\mu_{mc} B}{K_B + B} X_c \quad \text{(14.A4)}$$

When the mass balance around the secondary clarifier is included in order to compute X_{cu}, it is readily established that this system of equations exhibits the multiplicity of steady states and the possibility of *washout* described in Section 4.7; a minimum residence time is thus required.

The growth characteristics of the microorganisms are often expressed as functions of the *mean sludge age*, θ_c, which is defined as X_c/r_c; an alternative expression is

$$\theta_c = \frac{V_a X_c}{wQ_f X_{cu} + (1-w)Q_f X_{ce}} \quad (14.A5)$$

This is the ratio of organism mass holdup in the aerator to organism mass flow rate through the system; if it is assumed that most of the biomass is in the aerator, then the sludge age roughly equals the mean microorganism residence time in the system. The sludge age is roughly equal to $(r + w)/w$ times the aerator residence time, and the criterion for washout can also be expressed in terms of a minimum sludge age.

Certain nitrogen-containing species are also converted in a two-step process:

$$2NH_4^+ + 3O_2 \rightarrow 2NO_2^- + 2H_2O + 4H^+$$

$$2NO_2^- + O_2 \rightarrow 2NO_3^-$$

The first of these reactions is carried out by the bacterium *Nitrosomonas*, and the second by *Nitrobacter*. The mass balances and kinetics, with appropriate changes in subscripts, are identical to Eqs. (14.A1) through (14.A5). Typical literature values for the parameters for the three species are shown in Table 14.A1. The sludge age for all species in the reactor can be shown to be the same, so a minimum sludge age of at least four days will be required to prevent washout of the *Nitrosomonas*.

It is convenient to make one important kinetic assumption. BOD exists in the wastewater in both dissolved and suspended form. The volatile suspended solids (VSS) in the aerator thus consist of both organic waste and living biomass. The organic waste is a small fraction of the total and it is neglected insofar as solids transport is concerned; it is assumed for reactor calculations that the kinetic constants are the same for the suspended and

TABLE 14.A1
Typical kinetic constants for carbonaceous and nitrifying bacteria at 20°C

Constant	Carbonaceous	Nitrosomonas	Nitrobacter
μ_m	5.0 day^{-1}	0.33 day^{-1}	0.80 day^{-1}
b	0.055 day^{-1}	0.05 day^{-1}	0.05 day^{-1}
K	100 g/m^3 of BOD$_5$	1.0 g/m^3 of NH$_4$-N*	2.1 g/m^3 of NO$_2$-N*
Y	0.50 g VSS/g BOD$_5$	0.05 g VSS/g NH$_4$-N	0.02 g VSS/g NO$_2$-N
Minimum θ	0.2 days	3.6 days	1.3 days

*NH$_4$-N denotes "nitrogen as ammonia." NO$_2$-N denotes "nitrogen as NO$_2$."

dissolved BOD. The available data are inadequate to evaluate the error introduced by this assumption, but kinetic rates (including mass transfer limitations) are undoubtedly slower for the suspended BOD.

For a specified throughput and wastewater characteristics, there are two further degrees of freedom in order to size the aeration basin using the equations given here: Sludge age and reactor suspended solids (MLVSS, or *mixed liquor volatile suspended solids*) are the two best suited to overall plant design calculations, which then give the reactor size and effluent properties. For all designs discussed here, the width and depth of the rectangular aeration basins were taken to be 6 m, following a typical industry rule of thumb. The length of an individual basin was kept to no more than 45 m, which is definitely at the extreme limits of applicability of the perfect mixing assumption. Multiple reactors in parallel were used if a larger volume was required.

The air supply system to the aeration basin is a major factor in both capital and operating costs. The oxygen mass balance includes the utilization rate and the rate of oxygen transfer from the diffused air. The former is simply proportional to organism growth rates through a stoichiometric coefficient for each biological reaction; the latter equals $\alpha k_L a (DO-DO_{sat})V_a$, where DO and DO_{sat} are the actual and saturation dissolved oxygen levels, respectively, and $k_L a$ is the product of an average mass transfer coefficient and gas-liquid interfacial area/reactor volume. $k_L a$ is obtained from a correlation for tanks of similar size, where it is found to be linear in air flow rate/tank volume. α is a weak function of θ_c that adjusts mass transfer rates measured in water for use in wastewater. Thus, when the reactor has been sized, the air supply system requirement is immediately determined. For the base case design in Section 14.4, the computed value of 70.3 m^3 air/kg BOD is in the middle of the rule-of-thumb range of 30 to 110 m^3 air/kg BOD.

14.A2   CONTINUOUS SEDIMENTATION

14.A2.1   Introduction

The secondary clarifier, shown schematically in Fig. 14.A1, performs two functions in the activated sludge process: concentration of the microorganism suspension (*thickening*) for recycling and wastage, and *clarification* of the effluent for disposal. The feed enters through a radial flow diffuser, which disperses solids equally in all directions. Most of the sludge settles out and forms a concentrated *sludge blanket* at the bottom of the settler. The blanket is continuously raked to the center by a series of scraper blades and

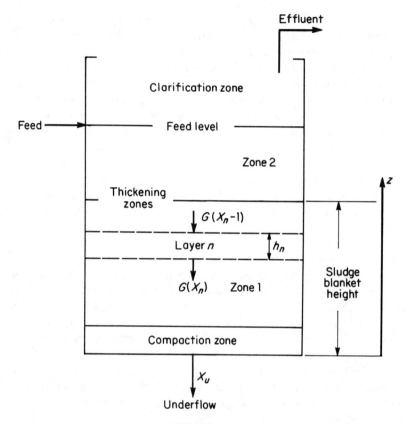

FIG. 14.A1.
Schematic of secondary clarifier. The layers are a consequence of flux theory, and some of the nomenclature is relevant only for the dynamic modeling.

pumped through the underflow line. Particles that do not settle are carried out over the effluent weir. The modeling of continuous thickening is reasonably well developed, and comprises most of this section. Clarification can be treated only empirically at present. The clarification function determines the volatile suspended solids (VSS) level in the effluent, which is one of the design standards.* In addition, the organic content of the suspended solids makes up a substantial fraction of the effluent BOD, which is the other design standard.

The analysis of continuous sedimentation is based on an approach

---

*Effluent criteria are specified on the basis of total suspended solids (TSS). Volatile suspended solids (VSS) typically comprise 80% of TSS, and this factor is kept constant throughout the modeling and simulation.

14.A2 CONTINUOUS SEDIMENTATION

commonly referred to as *flux theory*. The real validity of this approach is probably restricted to rigid monodisperse suspensions, and one often hears comments that it is not appropriate for activated sludge slurries. This criticism is undoubtedly overstated; application of the theory in the environmental engineering literature has been through graphical methods in which the estimate of a tangent and the width of a pencil point can cause errors of over 60% in a computed concentration. In any event, flux theory is the only approach to design and control of these units that is currently available.

14.A2.2 Flux Theory

Solids move to the bottom of a settler by two mechanisms: because there is a net convective flow of the slurry to the bottom, and because solids, being heavier, settle *relative* to the surrounding fluid. We will take the cross section to be cylindrical, and distance z measured from the bottom. Let $u = (r + w)Q_f/A_s$, where A_s is the settler cross section, and let X be the concentration of solids at any height. $v_s(X)$ is the velocity at which a particle settles relative to the surrounding fluid at solids concentration X; $v_s(X)$ is expected to be a decreasing function of concentration, since the settling becomes hindered as particles interact. The mass balance on the suspended phase is then

$$\frac{\partial}{\partial t} A_x \Delta z X = A_s uX|_{z+\Delta z} - A_s uX|_z + A_s v_s(X)X|_{z+\Delta z} - A_s v_s(X)X|_z$$

or, dividing by $A_s \, \Delta z$ and taking the limit as $\Delta z \to 0$, we obtain

$$\frac{\partial X}{\partial t} = \frac{\partial}{\partial z} uX + \frac{\partial}{\partial z} v_s(X)X \qquad (14.A6)$$

Equation (14.A6) is hyperbolic, which means that discontinuities can be propagated. In fact, the equations of continuous sedimentation are similar to those that arise in compressible gas dynamics, and shock and rarefaction waves like those in gas dynamics are expected to occur. It is important to note that the location of a discontinuity at any time is uniquely set by the conditions at $t = 0$; this is unlike the more familiar situation in which the final steady-state profile is independent of initial conditions. (The hyperbolic structure is a consequence of the fact that we have included only convective slurry and particle fluxes in Eq. (14.A6), and we have neglected any local dispersion that would have the appearance of a Brownian motionlike phenomenon (i.e., diffusion); the latter lead to a term proportional to $\partial^2 X/\partial z^2$, making the system parabolic. The parabolic system would have a unique steady-state concentration profile. Since the dispersion coefficient would

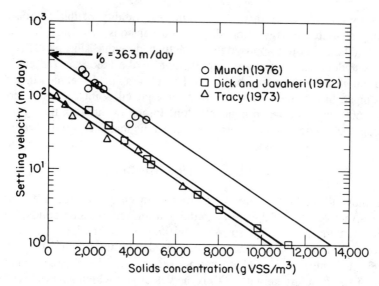

FIG. 14.A2.
Settling velocity as a function of concentration.

probably be quite small, the equation would be nearly singular, with narrow boundary layers replacing the abrupt shock waves, but with the locations of the steady-state near-discontinuities fixed independently of initial conditions.)

We now consider steady-state operation of the settler. Equation (14.A6) has the steady-state solution

$$G(X) = uX + v_s(X)X = \text{constant} \qquad (14.\text{A}7)$$

G is the flux (mass/area/time) of solids, which must be the same at each position in the settler. $v_s(X)$ is determined by batch experiments that are described in detail in environmental engineering textbooks. All the data of which we are aware on real activated sludge (three sets!) are plotted in Fig. 14.A2 on semilogarithmic coordinates; while the best line through each data set has a slightly different slope, statistical tests of significance establish that all are fit by the equation

$$v_s(X) = v_0 \exp(-4.5 \times 10^{-4} X) \qquad (14.\text{A}8)$$

with X measured in g/m^3 of VSS. As we shall discuss subsequently, v_0 is probably a function of sludge age.

The function $G(X)$ is plotted in Fig. 14.A3 for the best fit to the data of Munch and $u = 13.1$ m/day; the latter was the underflow velocity in some experiments to be discussed. The first observation to be made is that, for a specified flux, there are multiple steady-state solutions, corresponding to low concentration–high settling rate, high concentration–low settling rate, and an intermediate rate. Heuristic arguments exist purporting to show that steady states having fluxes higher than that given by the minimum in $G(X)$ are unstable, that the flux at the minimum represents the maximum rate at which solids can be conveyed, and hence this *limiting flux* represents the flux at which the settler must operate. These heuristic arguments are spurious, but a detailed analysis of the hyperbolic system shows that the system will, in nearly all cases, go to the limiting flux. (Bottom conditions can be constructed where this will not be the case.) We accept the limiting flux as the design basis, in which case we expect two concentration layers, with the lower concentration (lighter) layer on top. Attempts to operate with a higher flux will cause solids to accumulate, raising the height of the sludge blanket; this condition is known as *overloaded,* and if unchecked will increase the solids in the effluent. Lower fluxes will cause the system to operate under conditions where only a single, low-concentration steady-state layer exists; this *underloaded* condition affects dynamics adversely.

There is one final factor in the steady-state analysis. The concentrated slurry that is removed from the bottom of the thickening section must have a mass flow rate at steady state that is equal to the limiting flux, but there

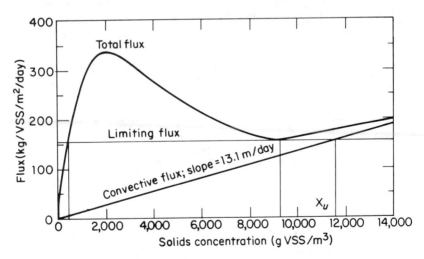

FIG. 14.A3.
Computed flux using data of Munch (1978), $u = 13.1$ m/day.
The flux curve differs somewhat from that computed using Eq.
(14.A8) and $v_0 = 363$ m/day.

is no sedimentation in this section. Thus, the underflow concentration, X_u, must simply be given by the equation

$$G(X_u) = uX_u \tag{14.A9}$$

This point corresponds to the intersection of the convective flux line (slope u) with the horizontal tangent corresponding to the limiting flux. We therefore expect to find three concentrations in the thickening section, corresponding to the dilute, concentrated, and underflow calculations. While the graphical construction is illustrative, it is obviously more convenient and more accurate to differentiate the flux equation and solve the resulting algebraic equations, which we do not reproduce here.

The model predictions are compared with an experiment of Munch at the Chicago Southwest Sewage Plant in Table 14.A2. The agreement between experimental and computed underflow concentration and limiting flux is quite good. (The reported underflow and limiting flux, which are taken from different sources, are not consistent for the given underflow velocity, but the difference is small.) As might be expected, the calculation is sensitive to details of the settling curve, and use of the best straight line through Munch's settling data alone [$v_s = 456 \exp(-5.32 \times 10^{-4}X)$] gives a limiting flux of 152 kg/m^2/day and an underflow concentration 11.6 kg/m^3. (Munch used a graphical construction to compute the limiting flux for a slightly different throughput and obtained 262 kg/m^2/day, while the numerical solution of the equations gives 155. It is this type of graphical construction that has contributed to the impression that the flux theory is inadequate.)

There are few data on concentration profiles. Anderson reported measurements in 1945 on a settler of the same diameter at the same Chicago Southwest Sewage Plant. While it is unlikely that the sludge settling characteristics remained unchanged for 30 years, settler concentrations were

TABLE 14.A2.
Model predictions compared to experiment for a settler with diameter 38.4 m and underflow velocity 13.1 m/day

	Model	Experiment (Munch, 1976, 1978)
Limiting flux (kg VSS/m^2/day)	171	160
Dilute zone concentration (kg VSS/m^3)	0.59	
Concentrated zone concentration (kg VSS/m^3)	10.2	
Underflow concentration (kg VSS/m^3)	13.1	13.6

14.A2 CONTINUOUS SEDIMENTATION

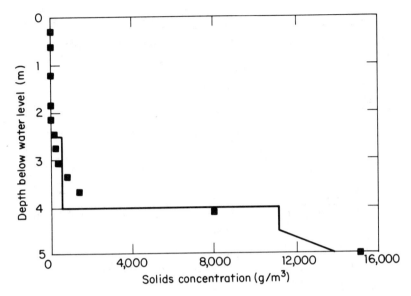

FIG. 14.A4.
Sludge concentration as a function of depth at the sampling point closest to the center, data of Anderson (1945) at Chicago Southwest Sewage Plant. Other sampling points show similar behavior.

computed using the same data for the experimental underflow of 9.8 m/day. The computed value of the limiting flux was 13.9 kg/m^3/day, while the experimental value was 15.1. Typical profile data are shown in Fig. 14.A4, together with the values computed for each zone using the flux equations. (It is to be recalled that only the value of the concentration can be computed, and not the position in the settler.) There is no sharp discontinuity as predicted by the theory, but the computed values seem quite reasonable. Both dispersion and a particle-size distribution could account for the smooth transition.

14.A2.3 Settling Velocity

Sludge settling is expected to depend on sludge age. It is believed that flocculation is enhanced by the presence of exocellular polymer secreted by the microorganisms; this secretion seems to occur at the greatest rate during endogenous growth, so improved ability to settle should be observed as the sludge age is increased. At very long sludge ages, on the other hand, very high rates of endogeneous respiration could cause cell lysis (cell breakup),

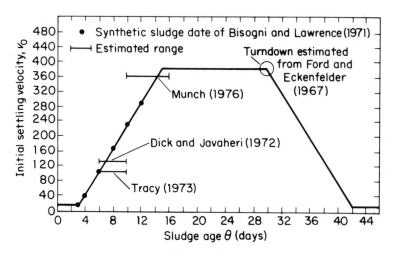

FIG. 14.A5.
v_0 as a function of sludge age (Paterson and Denn, 1980).

leading to floc breakup and poorer settling. The makeup of the cell population with sludge age is another important factor in settling characteristics. At low sludge ages the population contains a high proportion of long, rigid filamentous bacteria, which settle poorly. At high sludge ages the population contains a high proportion of zoogleal bacteria, which do not form flocs easily. A mixture of the two is required for good settling so that the filamentous bacteria can form a backbone to which the zoogleal bacteria attach.

Data showing the dependence of v_0 on sludge age are extremely limited. Measurements of Bisogni and Lawrence are shown in Fig. 14.A5; these data were taken on a sludge grown in a laboratory reactor on a synthetic substrate. (Sludge ages for the data in Fig. 14.A2 were not given, but were estimated from the known characteristics of the plants and are also shown in Fig. 14.A5. They are probably consistent with the laboratory data, but no stronger statement can be made.) The existence of a region of constant v_0 and the subsequent turndown in settling rate at $\theta_c = 30$ days are based on observations of Ford and Eckenfelder. The slope in the declining region is not known.

Design calculations usually assume a settling curve that is independent of sludge age. Simulations for the study described here were done for three cases: v_0 given by the function in Fig. 14.A5, by the extrapolation of the rising portion of the function in Fig. 14.A5, and by a constant value of 143 m/day.

14.A2.4 Clarification

There is no theoretical basis at present for estimating the solids carried over in the clarified effluent, and empirical results must be used. One correlation exists relating effluent suspended solids to feed flow rate and solids loading; food-to-microorganism ratio $(-r_B/X_C)$, which is closely related to sludge age; and settler depth. The correlation coefficient, however, is only 0.63, so it is of little use for design purposes. We have chosen to use a linear regression equation obtained from a single plant over a period of 41 days. Sludge age does not enter the equation, which is

$$X_e = 0.267 (1 - w) \frac{Q_f}{A_s} - 2.7 \times 10^{-3} X_a + 15 \qquad (14.\text{A}10)$$

with Q_f/A_s in m³/m²/day and X_e and X_a (effluent and aeration basin suspended solids, respectively) in g/m³. The multiple regression coefficient here was 0.92. While it is likely that θ_c will affect the effluent solids level, available data do not support its inclusion in the working equations.

The rule of thumb for secondary settlers is to design for an overflow of between 20 and 30 m³/m²/day. The effluent standards used in this work call for effluent TSS of not more than 20 g/m³, or effluent VSS (X_e) not more than 16 g/m³. A typical plant will operate with X_a in the neighborhood of 2000 g/m³. Using this number in Eq. (14.A10) gives an overflow rate of 24 m³/m²/day, which is consistent with standard practice. Extrapolation to extreme conditions, where sludge age could be a significant factor, would be a very risky thing to do, however, and it is clear that this is an area in which considerable research could profitably be done.

14.A2.5 Design

The first estimate of a settler area is computed from Eq. (14.A10) for specified effluent properties and aeration basin VSS. The settler depth is taken to be 5 m; the diameter is restricted to less than 40 m, based on industry practice. Multiple settlers are specified if more area is required. The actual diameter is set at the first integral value above the calculated diameter. The concentrations in the thickening section and the underflow are then computed using Eqs. (14.A7) through (14.A9), together with the limiting flux concept (that is, G is chosen such that $\partial G/\partial X = 0$).

If the computed flux is too large, then the unit must be sized with a larger area on the basis of thickening. The maximum flux that can be tolerated according to the flux theory will occur when the minimum in the flux curve, Fig. 14.A3, vanishes; this corresponds to the simultaneous vanishing of the first and second derivatives of $G(X)$, giving a value

$$G_{max} = \frac{4v_0}{4.5 \times 10^{-4}e^2} \simeq 1200\, v_0$$

If the area must be increased to allow thickening, then the effluent will, of course, be below the required solids level.

14.A2.6 Dynamics

A dynamical model allowing for arbitrary changes in feed characteristics and underflow rate cannot be obtained by numerical solution of Eq. (14.A6) using any reasonable algorithm for the solution of hyperbolic equations because of the complexity and sheer magnitude of the task. A number of models have been developed that attempt to approximate the important features of Eq. (14.A6), including, of course, the existence of discontinuities. The propagation of a discontinuity is therefore an important element of any model. Discontinuities propagate along the characteristic curves of the hyperbolic equation, and these can be written formally with no further development, but it is better to derive them directly.

Let Z_i be the height of the ith discontinuity. The concentration in the layer above Z_i is X_{i+1}, and the concentration in the layer below is X_i. Now consider a mass balance in a moving coordinate system extending a fixed distance Δz_{i+1} above Z_i and a fixed distance Δz_i below:

$$\frac{d}{dt}\int_{Z_i-\Delta z_i}^{Z_i+\Delta z_{i+1}} X(z)dz = G|_{Z_i+\Delta z_{i+1}} - G|_{Z_i-\Delta z_i}$$

X is a constant on each side of the discontinuity, so application of the Leibniz rule for differentiating integrals gives

$$X_{i+1}\frac{dZ_i}{dt} - X_i\frac{dZ_i}{dt} = G(X_{i+1}) - G(X_i)$$

or

$$\frac{dZ_i}{dt} = \frac{G(X_{i+1}) - G(X_i)}{X_{i+1} - X_i} \tag{14.A11}$$

Here, X_i denotes the solids level in layer i.

14.A2 CONTINUOUS SEDIMENTATION

The dynamic model assumes that the settler consists of four zones: (1) a well-mixed clarification zone, in which the effluent is determined even dynamically by Eq. (14.A10); (2) a zone containing layers of low solids concentration, the boundaries of which vary in time according to Eq. (14.A11), with the concentration of the uppermost layer possibly time dependent to account for time variations in the solids flux to the thickening section; (3) a dense zone containing layers in which the time-invariant concentrations correspond to the limiting concentration for some underflow velocity that the system has experienced in the past; and (4) a compaction zone of constant volume in which the concentration equals the underflow concentration. A layer is removed whenever its thickness as computed by Eq. (14.A11) goes to zero (that is, $Z_i \to Z_{i+1}$). A new layer may form whenever either the underflow rate or the feed flux is changed. In the former case a layer of infinitesimal thickness is created in the dense zone with a concentration equal to the limiting value for that underflow. The layer will grow in height as long as the solids flux entering the layer exceeds the limiting flux at that underflow rate; otherwise the layer will not develop. A new layer will form at the top of the dilute zone if the solids flux to the thickening section is less than the flux in the uppermost layer of the dilute zone.

Momentum conservation is included in the model only in one regard: adverse concentration gradients are known to be an unstable configuration which should induce mixing, so that, if at any time the concentration of a layer exceeds that of the layer below it, the two layers are combined at their average concentration.

One further technical detail must be included, known as Lax's generalized entropy condition and corresponding to the second law of thermodynamics in the gas dynamics analogy. A discontinuity between adjacent layers is not allowed if either of two conditions is satisfied:

$$\left.\frac{dG}{dX}\right|_{X=X_i} > \frac{dZ_i}{dt} \quad \text{or} \quad \left.\frac{dG}{dX}\right|_{X=X_i} < \frac{dZ_{i-1}}{dt} \qquad (14.A12)$$

In that case an approximation to a continuous variation in concentration (*a rarefaction wave*) must be included in the model.

Implementation of a computer code embodying the dynamic model is straightforward but does require some care with the logic. Since layers form and disappear, the number of differential equations changes with time, and indexing can be a problem. Each time a new layer is formed the entropy condition must be checked, and provisions (not discussed here) made to introduce a smooth transition if the condition is violated. In order to keep the number of layers from growing too large, concentration differences between adjacent layers are monitored, and layers are combined if the differences are less than some preset number.

14.A3 PRIMARY CLARIFIER

A fundamental model of the primary clarifier is not possible at this time. Such properties of the suspended solids in the wastewater as density and size distribution are generally unknown and changing in time, and complicated entrance and exit effects will in any event confound calculation based on particle sedimentation equations. Industry rules of thumb exist. The *Ten States Standards* call for a depth of 3 m and a holdup of about 2 hr for plants with an average flow exceeding 5600 m^3/day. Metcalf and Eddy suggest an area A_{ps} such that Q_f/A_{ps} is in the range 15 to 50 m^3/m^2/day, with a design basis of 50 m^3/m^2/day at peak loading for untreated wastewaters; this guideline is expected to result in removal of 50% to 60% of influent suspended solids.

Data are available on solids removal from a large number of primary clarifiers, as shown in Fig. 14.A6. There is a great deal of scatter, but the least-squares line does follow the general trend and is consistent with the 50 m^3/m^2/day rule to achieve 50% to 60% removal. The spread is narrowed somewhat if only rectangular settlers are included. The best available empirical model, based on multivariable linear regression, is therefore

$$FR = 0.64 - 3.8 \times 10^{-3} \frac{Q_f}{A_{ps}} + 2 \times 10^{-4} \, TSS_{pf} \qquad (14.A13)$$

FR is the fractional removal of solids, A_{ps} is the primary clarifier area, and TSS_{pf} is the total suspended solids in the feed to the primary clarifier. Limited data on the addition of anionic polymer to enhance settling suggest including a multiplicative factor of 1 plus 0.05 times the polymer concentration, and this is included in the model with the constraint that FR cannot exceed unity.

Equation (14.A13) is used to size the primary clarifier with a fixed depth of 3 m, a width of 6 m, and a maximum length based on industry practice of 30 m. At least two parallel units are always used to ensure that at least one unit is always available for partial treatment. The underflow is assumed to be thickened to 50,000 g TSS/m^3, which is typical of primary settlers in service. There was no attempt here to apply flux theory, since settling data for primary solids are not available.

Equation (14.A13) provides a design basis that is consistent with the rules of thumb. For a 20,000 m^3/day plant with 200 g TSS/m^3 in the feed, four clarifiers are required with an overflow rate of about 20 m^3/m^2/day and a holdup of 3.6 hr to achieve 60% solids removal. As noted in the body of this chapter, the primary clarifier costs have a major impact on the plant economics. A much better data base is therefore required for this unit in

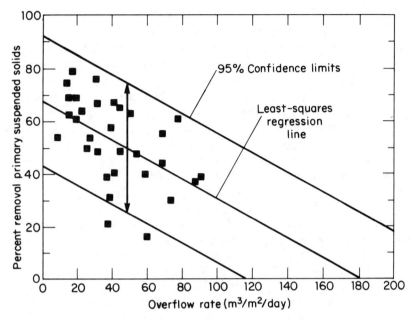

FIG. 14.A6.
Removal of suspended solids as a function of overflow rate in full-scale primary settlers (compiled by Water Pollution Control Federation, 1976; from Paterson and Denn, 1980).

order to develop a model that can truly be used with confidence for rational design.

14.A4 OTHER UNIT OPERATIONS

The preliminary treatment is simply screening and grit removal to remove large objects and high-density particles, and is not modeled. The chlorine contact basin was sized as a plug flow reactor with a 15-min residence time at the expected maximum flow rate ($1.4Q_f$), following the industry standard, with addition of 5 g Cl_2/m^3 effluent.

The sludge thickener functions essentially as does the secondary clarifier, but it is receiving a feed consisting of concentrated slurries from the primary and secondary underflows. Settling data for such a system do not exist, and the design equations were developed empirically using the following industry rules of thumb: For primary sludge, the loading rate is 98 kg TSS/m^2/day; for secondary sludge from the usual activated sludge process (θ_c = 5 to 10 days), the loading rate is 30 kg TSS/m^3/m^2/day; and for secondary sludge from the extended aeration process (θ_c = 10 to 20 days), the loading rate is up to a maximum of 88 kg TSS/m^2/day. The secondary sludge loading rate was therefore taken to be 20 kg/m^3/day for $\theta_c \leq 5$, 88 kg/m^2/day for $\theta_c \geq 20$, and to vary linearly between those limits for $5 < \theta_c < 20$; the total loading rate was then taken to be the weighted average of the primary and secondary values for the given feed rates. The maximum diameter permitted for a single thickener was 40 m. Underflow was assumed to be thickened to 90 kg TSS/m^3, which is the normally expected value. The effluent solids concentration is taken to be zero, since it will be too small to have any influence on reactor design or performance.

The anaerobic digester is a biochemical reactor in which anaerobic microorganisms break down the organic solids in the thickened sludge. This is typically a two-stage system, in which the first stage is a well-stirred reactor and the second is used mainly for separation. Equations (14.A1) through (14.A5), with suitably altered subscripts and $r = 0$, apply to the reactor. The following published kinetic parameters were used for operation at 35° C: $\mu_m = 0.28$ day^{-1}, $b = 0.023$ day^{-1}, $K = 154$ g/m^3 of BOD$_5$, and $Y = 0.04$ g VSS/BOD$_5$. The design basis was 60% reduction in solids; this was found to give volumes considerably below the standard loading rate of 480–1600 g VSS/m^3/day, and a safety factor of four (!) was used in the actual sizing of the first stage in order to conform to the industry standard. The second stage was taken to be equal in size to the first, with an underflow of 80 kg TSS/m^3 and an effluent containing no BOD or solids.

The solids centrifuge was assumed to operate at 93% solids capture for a feed rate of 87 kg TSS/hr, based on published values, with concentration to 200 kg TSS/m^3.

BIBLIOGRAPHICAL NOTES

The standard texts in this area are

Metcalf and Eddy Inc., *Wastewater Engineering* (New York, NY: McGraw-Hill, 1972).

R. S. Ramalho, *Introduction to Wastewater Treatment Processes* (New York, NY: Academic Press, 1977).

There is an excellent review in

D. Taylor, Y. Smeers, and E. Nyns, CRC Critical Reviews in Environmental Control (1972).

The chapter is largely based on

> R. Paterson and M. M. Denn, "Design and Control of an Activated Sludge Process for Municipal Wastewater Treatment," *Technical Completion Report, A-041-DEL,* Water Resources Center, University of Delaware, Newark, DE, 1980.

The report is summarized in part in

> R. B. Paterson and M. M. Denn, *Chem. Eng. J.* **27** (1983) B13.

The most illuminating treatment of the dynamics of continuous sedimentation is in

> C. A. Petty, *Chem. Eng. Science* **30** (1975) 1451.

The dynamical model used here is described in

> U. Attir, M. M. Denn, and C. A. Petty, *AIChE Symp. Series* **73**, No. 167 (1977), 49.

The dynamics of recycle processes is addressed in

> M. M. Denn and R. Lavie, *Chem. Eng. J.* **24** (1982) 55.

The first discussion seems to be provided by

> E. R. Gilliland, L. A. Gould, and T. H. Boyle, *Preprints 1964 Joint Automatic Control Conference,* Stanford University (1964).

(These JACC Preprints were usually widely distributed and subscribed to by many libraries.) The treatment in the present context (and insofar as we know, the only other paper on the subject) is in

> U. Attir and M. M. Denn, *AIChE J.* **24** (1978) 693.

A FINAL COMMENT

A few closing remarks are appropriate here. The treatment of modeling has obviously been selective, but the important concepts should be clear; so too should their applicability to situations far beyond the bounds set by the examples considered here.

Both mathematical analysis and numerical computation have been assigned a minor role in our treatment (though some readers might question this evaluation); this is a consequence of the belief that these handmaidens have too often become the focus, receiving an undue emphasis that obscures the more basic problem of determining the equations to be solved. Computation and mathematical analysis are important in applications, but they are not the essence of modeling!

Finally, it is obvious that the material covered in this text is just the beginning. The complexity of most processes requires more empiricism than we have utilized in our examples, and the mathematical structures are often far more complex than we have encountered here. Yet the basic ideas remain unchanged, and the guidance of a good fundamental model even in the development of empiricisms and heuristics is invaluable; for many situations a good fundamental model is all that is needed for the insight to solve an engineering problem.

AUTHOR INDEX

Achinstein, P. 11
Ahner, D. J. 161
Aiken, R. C. 204
Akhiezer, N. I. 217
Amundson, N. R. 204
Araki, M. 269
Aris, R. 10, 44, 98, 100, 124
Attir, U. 319
Avenas, P. 267

Baillagou, P. 229
Bamford, C. H. 229
Becker, E. B. 217
Bird, R. B. 98, 124
Bliss, G. A. 217
Bogue, D. C. 268
Bonvin, D. 204
Boyle, T. H. 319
Brower, A. S. 161

Carey, G. F. 217
Carroll, L. 275
Cartwright, N. 11
Chang, J. C. 268, 269
Cheung, Y. K. 218
Chimowitz, E. H. 204
Coyle, D. J. 229
Crochet, M. J. 218, 267
Cussler, E. L. 98
Cwiklinski, R. 161

Davies, A. R. 218
Davis, S. H. 267
Dealy, J. M. 267
Denn, M. M. 30, 43, 98, 99, 124, 160, 161, 177, 204, 217, 267, 268, 269, 318
Dewes, M. H. 161
Duhem, P. 11

Eckhaus, W. 204
Eng, M. T. 101
Evans, L. B. 101

Fan, L. T. 99
Finlayson, B. A. 217
Fisher, R. J. 204, 267, 268
Franks, R. G. E. 99
Friedly, J. C. 98, 204

Gagon, D. K. 268
George, H. H. 30, 268
Gilliland, E. R. 319
Gould, L. A. 98, 101, 204, 228, 319
Gray, B. R. 100
Gray, R. 100
Greenbaum, B. 204
Gross, B. 102

Hahn, W. 177
Hesse, M. B. 10, 11, 275, 276
Hlavacek, V. 100, 101
Hulbert, H. M. 229

Isaacs, B. 101
Iscol, L. 102
Ishihara, H. 30

Jacob, S. M. 102
Jackson, R. F. 228
Jensen, K. F. 99

Kahlert, C. 100
Kase, S. 30, 268, 269
Katz, S. 229
Keunings, R. 267
Kirwan, N. A. 100
Krishna, R. P. 99
Kugelman, A. M. 102

Kuhn, T. S. 276
Kurihara, H. 101

Lamb, D. E. 101
Lapidus, L. 204
LaSalle, J. 177
Lavie, R. 319
Lee, W. 101, 102
Lefschetz, S. 177
Lewin, D. R. 204
Liapunov, A. A. 177
Lightfoot, E. N. 98, 124
Lin, C. C. 124
Lin, S. H. 99
Lorenz, E. N. 100
Lu, C. H. 161
Luss, D. 204
Luyben, W. L. 101

Makila, P. M. 98
Matovich, M. A. 267
May, R. M. 100
Mellichamp, D. A. 204
Metcalf and Eddy Inc. 318
Michelsen, M. L. 217

Nakano, N. 101
Nam, S. 268
Nyns, E. 318

Oden, J. T. 217
Oistrach, S. 124

Pars, L. A. 217
Patel, A. S. 161
Paterson, R. B. 319
Pearson, J. R. A. 267, 268
Perlmutter, D. D. 177
Petrie, C. J. S. 99, 267, 268
Petty, C. A. 319
Pigford, R. L. 228
Poore, A. 99
Popper, K. R. 276

Ramalho, R. S. 318
Rathousky, J. 101
Ray, W. H. 99, 229

Redhead, M. 11
Rosenbrock, H. H. 228
Rossler, O. 100
Ruckenstein, E. 124
Ruelle, D. 100
Russell, T. W. F. 30, 43, 44, 98
Ruszynski, J. 101

Saidel, G. M. 229
Schmitz, R. A. 100
Schultz, W. W. 267
Segal, L. A. 124
Shinnar, R. 101, 160
Smeers, Y. 318
Soong, D. S. 229
Stewart, W. E. 98, 124
Stillman, R. 99
Storey, C. 228
Subramaniam, B. 100

Tanner, R. I. 267
Taylor, D. 318
Tirrell, M. 229
Tompa, H. 229
Toulmin, S. 276
Tulig, T. J. 229

Uppal, A. 99

Varma, A. 100
Villadsen, J. 217
Voltz, S. E. 102
Votruba, J. 100

Waller, K. V. 98
Walters, K. 218
Weekman Jr., V. W. 101, 102
Wei, J. 160, 161
White, J. L. 267

Yeo, M. F. 218
Yoon, H. 160, 161
Yu, W. C. 161

Zeichner, G. R. 268
Ziabicki, A. 267
Zienkiewicz, O. C. 218

SUBJECT INDEX

Activated sludge 27, 58, 277
Adjoint 181, 209
Aeration basin 302
Aeration tank 280

Birefringence 230
Biot number 108
BOD 283, 302
Bode diagram 5, 14, 162, 239, 266
Boundary layer 113, 140

Calculus of variations 211
Chaos 82
Clarifier 285, 305, 316
Coal gasification 17, 111, 125, 271
Collocation 149, 156, 207
Computation 148, 156
Conservation principles 32
Constitutive equations 9, 36, 38, 236
Continuous approximation 219, 223
Continuous drawing 24
Control 85, 94, 290, 297, 301
Control, closed loop 187
Control volume 34, 40
Controllability 288

Design 49, 277, 285, 300, 313
Diffusion equation 107
Dimensional analysis 123
Discretization 205
Distillation 220
Drag coefficient 117, 235, 240
Draw ratio 244, 260
Dynamics 290, 295, 314

Economics 280
Einstein diffusion time 111, 137
Einstein length 111
Energy 103

Energy, conservation of 62
Energy, internal 63, 235
Energy method 170
Enthalpy 63, 67, 235
Euler equation 212

Falsifiability, Degree of 274
Fiber spinline 24, 111, 272
Fiber spinning 230
Finite element 149, 205, 215, 246
Fluid catalytic cracker (FCC) 15, 84, 185
Fourier time 111
Frequency response 5

Galerkin method 207
Griffith number 118

Heat transfer 70, 71, 106, 107, 118, 150, 246
Heat transfer coefficient 235, 240

Infinitesimal perturbations 166
Instability 53, 78, 162, 169, 231, 260

Kinetics-free modeling 131, 271

Liapunov function 163, 169
Limit cycle 81
Linearization 196
Lubrication approximation 232

Mass, conservation of 45
Mass-spring dashpot system 2, 7, 9, 12, 165
Mathematical model 1, 12, 31
Maxwell model 238
Modal analysis 178, 184
Model, analogy 2, 5, 7
Model, empirical 2, 4
Model, fundamental 2, 3

323

Model, linear 5
Multiplicity 53, 74, 90

Navier-Stokes equations 113
Net present worth 281
Nussult number 108, 119, 240

Occam's razor 272
Order reduction 73, 93, 183

Partial molar quantities 42, 65
Perfect mixing assumption 34
Polymerization 225
Prandtl number 119

Reaction, autocatalytic 57
Reaction rate 48
Reactor, batch 60, 71
Reactor, biochemical 302
Reactor, coal gasification 125
Reactor, continuous-flow stirred 45, 167, 169
Reactor, fluidized bed 194, 196
Reactor, moving bed 136
Reactor, tubular 60, 71
Recycle 290
Rheology 236, 259
Ritz-Galerkin method 214

Scaling 103, 131
Sedimentation 305

Sensitivity 122, 158, 265, 270, 285, 296
Settler 280, 305
Simplicity 272, 274
Sludge age 303
Sludge blanket 290
Solidification 246
Stability 162
Stability, finite 172
Stability limits 196
Stoichiometric analysis 126
Strange attractors 83
Stress 236

Thin filament equations 231
Time constants 52
Toves 270
Transient 52, 81, 156, 231

Validation 90, 158, 270
Variables, characterizing dependent 32
Variables, fundamental dependent 32
Variables, independent 31
Viscoelasticity 237, 254, 273

Wastewater treatment 27, 58, 277
Weighted residuals 205
Well-stirred tank 32